Lung Mechanics

With mathematical and computational models furthering our understanding of lung mechanics, function, and disease, this book provides an all-inclusive introduction to the topic from a quantitative standpoint. Focusing on inverse modeling, the reader is guided through the theory in a logical progression, from the simplest models up to state-of-the-art models that are both dynamic and nonlinear.

Key tools used in biomedical engineering research, such as regression theory, linear and nonlinear systems theory, and the Fourier transform, are explained. Derivations of important physical relationships, such as the Poiseuille equation and the wave speed equation, from first principles are also provided. Examples of applications to experimental data illustrate physiological relevance throughout, whilst problem sets at the end of each chapter provide practice and test reader comprehension. This book is ideal for biomedical engineering and biophysics graduate students and researchers wishing to understand this emerging field.

Jason H. T. Bates is currently a Professor of Medicine and Molecular Physiology and Biophysics at the University of Vermont College of Medicine, and a Member of the Pulmonary Division at Fletcher Allen Health Care. He is also a Member of the American Physiological Society, the American Thoracic Society, and the Biomedical Engineering Society, and an elected Senior Member of the IEEE Engineering in Medicine and Biology Society. Dr. Bates has published more than 190 peer-reviewed journal papers in addition to numerous book chapters, conference abstracts, and other articles. In 1994 he was awarded the Doctor of Science degree by Canterbury University, New Zealand, and in 2002 he was elected a Fellow of the American Institute for Medical and Biological Engineering.

Lung Mechanics

An Inverse Modeling Approach

JASON H. T. BATES

University of Vermont

CAMBRIDGE
UNIVERSITY PRESS

CAMBRIDGE
UNIVERSITY PRESS

Shaftesbury Road, Cambridge CB2 8EA, United Kingdom

One Liberty Plaza, 20th Floor, New York, NY 10006, USA

477 Williamstown Road, Port Melbourne, VIC 3207, Australia

314–321, 3rd Floor, Plot 3, Splendor Forum, Jasola District Centre, New Delhi – 110025, India

103 Penang Road, #05–06/07, Visioncrest Commercial, Singapore 238467

Cambridge University Press is part of Cambridge University Press & Assessment, a department of the University of Cambridge.

We share the University's mission to contribute to society through the pursuit of education, learning and research at the highest international levels of excellence.

www.cambridge.org
Information on this title: www.cambridge.org/9780521509602

First published 2009

A catalogue record for this publication is available from the British Library

Library of Congress Cataloging-in-Publication data
Bates, Jason H. T.
Lung mechanics / Jason H. T. Bates.
     p.   cm.
Includes bibliographical references and index.
ISBN 978-0-521-50960-2 (hardback)
1. Lungs.   2. Human mechanics.   3. Respiration.   I. Title.
QP121.B35   2009
611'.24 – dc22   2009011365

ISBN   978-0-521-50960-2   Hardback

Dedicated with love to my wife, Nancy MacGregor, for her constant support.

Contents

Preface

Viewing the lungs as a mechanical system has intrigued engineers, physicists, and mathematicians for decades. Indeed, the field of lung mechanics is now mature and highly quantitative, making wide use of sophisticated mathematical and computational methods. Nevertheless, most books on lung mechanics are aimed primarily at physiologists and medical professionals, and are therefore somewhat lacking in the mathematical treatment necessary for a rigorous scientific introduction to the subject. This book attempts to fill that gap. Accordingly, some familiarity with the methods of applied mathematics, including basic calculus and differential equations, is assumed. The material covered is suitable for a first-year graduate course in bioengineering. I hope, however, it will also be accessible to motivated biologists and physiologists.

This book focuses on inverse models of lung mechanics, and is organized around the principle that these models can be arranged in a hierarchy of complexity. Chapter 1 expands on this concept and introduces the adjunct notion of forward modeling. It also sets the scene with a brief overview of pulmonary physiology in general. Chapter 2 attends to the fact that all the mathematical modeling skill in the world is for nought without good experimental data. Accordingly, this chapter is devoted to the key experimental methodologies that have provided the data on which the models described in subsequent chapters are based. It can thus be skipped without loss of continuity and referred back to when issues related to experimental validation of models arise. The discussion of inverse lung models begins in earnest in Chapter 3, which develops the theory behind the simplest plausible physiological model of all – a single elastic compartment served by a single flow-resistive airway. This represents the most basic level of inverse-model complexity, but one which still has a very useful physiological interpretation, discussed in Chapter 4. In proceeding to the second level of model complexity, we have a choice to make; is it more appropriate to require the elements of the simple model to be nonlinear, or should we add a second linear compartment? There is no simple answer to this question, so we proceed by examining nonlinear extensions of the basic model in Chapter 5 and go specifically into the nonlinear phenomenon of expiratory flow limitation in Chapter 6. The alternative to introducing nonlinearity, namely adding a second linear compartment, is developed in Chapter 7. This segues into the third level of complexity represented by the general linear dynamic model and the concept of impedance, discussed in Chapter 8. Various models of lung impedance are discussed in Chapter 9, while Chapter 10 is devoted to a particular example currently in widespread use, known as the constant phase model. Chapter 11 deals with the fourth and final level of complexity, that of the nonlinear dynamic model. Chapter 12 concludes the book with a brief overview in which the various inverse models considered in previous chapters are brought together into a unified picture.

In addition to covering the field of lung mechanics, this book also has a second goal. This is to exemplify how quantitative methods from the physical sciences can be used to advance knowledge in a biomedical subject of significant practical importance for human health. Lung mechanics is a prime example of this because it is so well suited to quantitative investigation, being essentially a manifestation of classical Newtonian physics in the body. However, there are many other areas of biomedical research that benefit from use of the same methods, which therefore have wide applicability. Accordingly, significant attention is paid to the explanation of these methods, which include multiple linear regression and its recursive formulation, statistical tests of model order, linear and nonlinear system identification, and the Fourier transform. Also, when the physiological interpretation of lung models is discussed, formulae encapsulating relevant physical processes are derived from first principles where possible. This is to try to minimize the uncomfortable sense of mystery that inevitably arises when any mathematical formula has to be taken on trust.

The material presented in this book stems from work carried out in numerous laboratories around the world, as well as research from my own laboratory over the past 25 years both at the Meakins-Christie Laboratories of McGill University and subsequently at the Vermont Lung Center of the University of Vermont College of Medicine. Some of what appears is thus the result of interactions I have had the privilege to enjoy with countless mentors, colleagues, and students. Space does not permit me to list everyone, much as I would like to. However, several friends and associates graciously read through drafts of this book and gave me their invaluable comments. On this account, my thanks go to (in alphabetical order) Gil Allen, Sharon Bullimore, Anne Dixon, Charlie Irvin, David Kaminsky, Anne-Marie Lauzon, and Bela Suki.

Notation

A	1) parameter in exponential pressure-volume relationship of lung
	2) area
	3) amplitude
A_0	equilibrium area of elastic tube
A_i	1) coefficient of exponential term
	2) coefficient of term in general second-order equation of motion
$\underline{A}$	vector of parameter values
$\underline{\hat{A}}$	estimate of parameter vector
A-D	analog to digital
AICc	corrected Akaike criterion
ΔA_i	change in the ith parameter value
a	parameter in sigmoidal pressure-volume relationship of lung
a_i	parameters in a series
B	parameter of exponential pressure-volume relationship of lung
B_i	coefficient of general second-order equation of motion
b	parameter in sigmoidal pressure-volume relationship of lung
b_i	parameter in a general linear differential equation
C	constant of integration
$C_{P,\dot{V}}$	cross-spectral density between pressure and flow
$C_{P,P}$	auto-spectral density of pressure
CD	coefficient of determination
c	parameter in sigmoidal pressure-volume relationship of lung
D	length of dashpot in tissue model
E	elastance
E_A	elastance of lung region under an alveolar capsule
E_{cw}	chest wall elastance
E_L	lung elastance
E_{rs}	respiratory system elastance
E_t	elastance of lung tissue
E_0	elastance of homogeneous lung model
E_1	volume-independent term in P_{el}
E_2	volume-dependent term in P_{el}
$E_i, i = 1,2\dots$	elastance of ith compartment
ΔE	increase in elastance due to mechanical heterogeneity

$F\{\}$	operation of Fourier transformation on bracketed quantity
f	frequency (Hz)
$f(x)$	general nonlinear function of x
FEV_1	volume of air exhaled in the first second of a forced expiration
FVC	forced vital capacity
G	tissue damping in the constant phase model
H	tissue elastance in the constant phase model
H	Heaviside step function
H_{min}	minimum value of H in the distributed constant phase model
H_{max}	maximum value of H in the distributed constant phase model
h	impulse response function
h_i	kernels in the Volterra series
I	inertance
I_{aw}	inertance of airway gas
I_t	inertance of lung tissue
$\underline{\underline{I}}$	identity matrix
J	cost function for Poiseuille flow formula
K	parameter in exponential pressure-volume relationship of lung
K_1	flow-independent term of the Rohrer equation for flow resistance
K_2	flow-dependent term of the Rohrer equation for flow resistance
k	1) spring constant in model of lung tissue strip
	2) power-law exponent
L	total length of model of lung tissue strip
l	length
M	mass
$\underline{\underline{M}}$	covariance matrix for recursive multiple linear regression
MSR	mean squared residual
m	number of model parameters
N	1) distribution of string lengths in model of lung tissue strip
	2) number of series springs in tissue model
$\underline{\underline{N}}$	vector of noise values in dependent variable
n	number of data points
P	pressure
$\bar{p}$	mean pressure
$\underline{P}$	vector of pressure measurements
P_A	alveolar pressure
P_{ao}	airway opening pressure
P_b	Bernoulli pressure
P_{box}	plethysmographic pressure
P_{el}	elastic pressure
P_{es}	esophageal pressure
$P_{impulse}$	impulse response in pressure
P_j	pressure at airway junction
P_{pl}	pleural pressure

P_{step}	step response in pressure
P_{tm}	airway transmural pressure
P_{tp}	transpulmonary pressure
P_0	1) baseline pressure
	2) initial pressure
P_i	pressure in ith compartment
PEEP	positive end-expiratory pressure
ΔP	change in pressure
ΔP_d	pressure drop across daughter airway
ΔP_{dif}	pressure change due to stress adaptation
ΔP_{init}	initial pressure change following flow interruption
ΔP_p	pressure drop across parent airway
p	dummy variable of integration
$\underline{\underline{Q}}$	matrix of noise values in independent variables
R	1) resistance
	2) real part of impedance
R_A	airway resistance leading into the lung from under an alveolar capsule
R_{aw}	airway resistance
R_c	resistance of common airway
R_{cw}	chest wall resistance
R_d	resistance of daughter airway
R_g	real part of thoracic gas impedance
R_{hole}	resistance of hole in pleural surface
R_L	lung resistance
R_N	Newtonian resistance of the constant phase model
R_p	resistance of parent airway
R_{rs}	respiratory system resistance
R_t	tissue resistance
R_0	resistance of homogeneous lung model
$R_i, i = 1, 2 \ldots$	resistance of ith compartment
Re	Reynolds number
ΔR	increase in resistance due to mechanical heterogeneity
r	radius
r_0	equilibrium radius of elastic tube
S	stress in viscoelastic tissue model
SSR	sum of squared residuals
T	tension
t	time
Δt	time step
u	dummy variable of integration representing time
V	volume
V_i	volume of ith compartment

υ	1) velocity
	2) voltage
$\dot{V}$	flow
$\dot{V}_{max}$	maximal expiratory flow
$\ddot{V}$	rate of change of flow
ΔV	change in volume
V_{TG}	thoracic gas volume
$\dot{W}$	rate of energy dissipation in laminar flow of fluid
X	imaginary part of impedance
X_{aw}	imaginary part of airway impedance
X_g	imaginary part of thoracic gas impedance
X_t	imaginary part of tissue impedance
$\underline{\underline{X}}$	matrix of independent variables
x	extension of spring in Maxwell body
Z	impedance
Z_{aw}	impedance of airways
Z_g	impedance of thoracic gas
Z_{in}	input impedance
Z_t	impedance of tissues
Z_{tr}	transfer impedance
Φ	coherence
α	1) Womersley number
	2) coefficient of exponential force-length relationship of lung tissue
	3) width of axial strip of airway wall
	4) sinusoidal coefficient
	5) exponent of frequency in the constant phase model
	6) exponent of elastic force in tissue
β	1) coefficient of exponential force-length relationship of lung tissue
	2) sinusoidal coefficient
	3) exponent of resistive force in tissue model
δ	1) asymmetry index in orders of the airway tree
	2) Dirac delta-function
ε	strain
λ	forgetting factor for recursive algorithms
σ	stress
σ^2	variance
$\hat{\sigma}^2$	estimate of variance
ϕ	phase
η	hysteresivity
μ	viscosity
ρ	density
τ	time-constant
ω	angular frequency

1   Introduction

1.1   The importance of lung mechanics

Being able to breathe without any apparent difficulty is something that healthy people take for granted, and most of us generally go about our daily lives without giving it a second thought. But breathing is not always easy. A number of common lung diseases can make breathing difficult and uncomfortable. Sometimes these diseases can even make it impossible to breathe at all without the assistance of a machine or another person, a condition known as *respiratory failure*. There are a variety of factors that can lead to respiratory insufficiency or failure, but among the most important are those that involve a compromise in the mechanical properties of the lungs.

Breathing is essentially a mechanical process in which the muscles of the thorax and abdomen, working together under the control of the brain, produce the pressures required to expand the lung so that air is sucked into it from the environment. These pressures must be sufficient to overcome the tendencies of the lung and chest wall tissues to recoil, much like blowing up a balloon. Pressure is also required to drive air along the pulmonary airways, a system of branching conduits that begins at the mouth and ends deep in the lungs at the point where air and blood are close enough to exchange oxygen and carbon dioxide. The mechanical properties of the lungs thus determine how muscular pressures, airway flows, and lung volumes are related. The field of lung mechanics is concerned with the study of these properties.

The mechanical properties of the lung have an important bearing on how we experience our daily lives because they determine, for example, how much effort is needed to take in a breath and how comfortable it feels to breathe. When breathing becomes uncomfortable, usually perceived as a sense of breathlessness known as *dyspnea*, our brains are telling us that we are expending too much effort to do what is normally effortless. In other words, we are sensing that there is something wrong with the mechanical properties of our lungs. This sensation can be reproduced by trying to breathe through a narrow drinking straw which presents a large resistance to air flow. A somewhat similar sensation may be experienced by someone suffering an attack of asthma, when the pulmonary airways constrict and so partially obstruct the flow of air into and out of the lungs. Taking a breath may also not be so easy when the lungs become encased in thick scar tissue, as occurs in a disease known as *pulmonary fibrosis*, somewhat like trying to breathe while wearing a tightly laced corset. But if pathologic abnormalities in lung mechanics are sensible to us as individuals, then they are also measurable using laboratory equipment. Indeed, physicians regularly assess mechanical abnormalities in the lung in order to diagnose disease. Assessing lung mechanical function is also vital to areas of basic science such as pulmonary pharmacology and immunology.

A great deal is known about lung mechanical function, thanks to the ongoing efforts of a large community of scientists dating back over 100 years. In its formative stages, beginning in the late 1800s and continuing throughout much of the twentieth century, the science of lung mechanics was largely the domain of the physiologist and physician. Over the past several decades, however, the field has progressed to become highly quantitative thanks to the availability of electronic sensors and digital computers. These devices have allowed investigators to acquire extremely accurate experimental data related to lung function. Accurate data are always very interesting to scientists armed with sophisticated methods of data analysis, so the field of lung mechanics has recently been attracting the attention of biomedical engineers, physicists, and mathematicians in increasing numbers. Indeed, we are now at the point where mathematics and computer models play indispensable roles in encapsulating our understanding of lung mechanics.

The field of lung mechanics thus represents a confluence of the biological and physical sciences, and as such requires a multidisciplinary approach in which questions of physiologic function are addressed in terms of underlying physics. The language of physics is mathematics, and its goal is to capture the workings of the world in terms of (ideally, relatively simple) equations that have broad predictive power. Accordingly, the approach taken herein is to systematically develop the equations (mathematical models) that describe lung mechanical function.

1.2   Anatomy and physiology

The physiological aspects of the lung can be rather naturally grouped into a number of almost distinct sub-topics: *gas exchange*, *neural control*, *mechanics*, and *non-respiratory functions* related mostly to defense. Indeed, advanced treatises on the lung invariably partition the subject along these lines, and even the corresponding communities of scientists currently pushing forward the frontiers of knowledge in these various areas remain largely distinct. We are not going to cross these boundaries to any significant degree here, being almost exclusively concerned with lung mechanical function. Nevertheless, it must be remembered that all aspects of pulmonary physiology are vital to the lung's ability to function normally within a human or animal, and to sustain life.

1.2.1   Gas exchange

Living animal cells require a continual supply of oxygen and nutrients, while continually releasing carbon dioxide and other waste products. Single-cell animals can achieve this through direct diffusive exchange with the environment. In larger animals, the increased volume-to-surface area ratios make it impossible for the necessary flux of gases between cells and the environment to be achieved by *passive diffusion* across the body surface. To deal with this problem, nature has evolved the *cardio-pulmonary system*, an intermediary

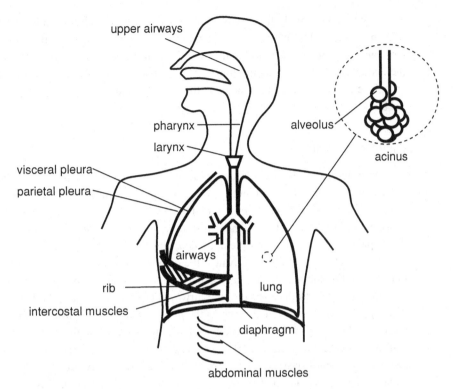

Figure 1.1 The principal mechanical components of the respiratory system. In a spontaneously breathing subject, a negative pressure is generated around the outside of the lungs by the respiratory muscles. This produces a flow of gas along the pulmonary airways in the direction of decreasing pressure.

that brings blood and gas into close juxtaposition either side of an extremely thin physical barrier deep inside the lungs. This *blood-gas barrier* is less than a micron thick, so it presents a very small impediment to the passage of gas molecules. It also has an extremely large surface area so that many gas molecules can cross it in parallel.

The enormous surface area of the blood-gas barrier is achieved by tree-like structures that geometrically amplify their cross-sections as their branches divide and become increasingly numerous. In the case of the *pulmonary airways* (Fig. 1.1), a cross-sectional area of a few square centimeters at the trunk of the tree (the *trachea*) is translated through about 23 bifurcations into an area roughly the size of a tennis court by the time the *alveoli* have been reached at the end of the most distal airway branches. A corresponding branching scheme begins with the pulmonary artery as it exits the right ventricle of the heart, and eventually leads to the myriad of pulmonary capillaries that distribute blood throughout the alveolar walls.

The transport of oxygen and carbon dioxide across the blood-gas barrier occurs solely by passive diffusion. Gases always tend to move from regions of high partial pressure to regions of low partial pressure, provided they are not physically prevented from doing so. The partial pressures of oxygen and carbon dioxide in the alveoli and the pulmonary capillary blood normally favor movement of oxygen into the blood and carbon dioxide into the alveoli. The blood-gas barrier these gases must cross in the process is so large and so thin that sufficient numbers of gas molecules can move across it to meet the demands of life in a healthy lung. In some diseases, however, this ceases to be the case. The efficiency of gas exchange is thus tightly linked to the physical properties of the blood-gas barrier.

1.2.2   Control of breathing

Gas exchange can only take place effectively if fresh air is continually supplied to the blood-gas barrier. This is achieved through the repeated inflation and deflation of the lungs under the coordinated actions of the *respiratory muscles*, a process known as ventilation. The principal muscles of inspiration are the *diaphragm*, a sheet of muscle that separates the thorax from the abdomen, and the *intercostal muscles* that stabilize the rib cage (Fig. 1.1). When activated, the diaphragm descends and, together with the muscles of the thorax, creates a negative pressure (relative to atmospheric) around the lungs that acts to draw air into the airways. The region around the lungs in which this negative pressure acts is known as the *pleural space*, and is filled with a very thin layer of lubricating fluid that separates the outer surface of the lungs (the *visceral pleura*) from the inner surface of the rib cage (the *parietal pleura*). Expiration under resting conditions is passive; the inspiratory muscles are deactivated to allow the lungs to deflate as a result of the net elastic recoil of the lung and chest wall tissues. The increased ventilatory demands of exercise may require the use of expiratory muscles, notably those of the abdomen, to increase the rate of expiration above that produced by elastic recoil alone.

The volume of air taken into the lungs with each breath, termed the *tidal volume* (Fig. 1.2), is substantially less than the total volume that can be forcibly expired from a maximal inspiration. This total volume, called *vital capacity*, is equal to the difference between *total lung capacity* and *residual volume*, the latter defined as the volume of air left in the lungs after a maximum expiratory effort. Residual volume is substantially less than the volume of air in the lungs at the end of a normal passive expiration, termed *functional residual capacity*.

Obviously, respiration requires that the various respiratory muscles be activated in a periodic and coordinated fashion. This is the job of the *respiratory control centers* in the brainstem, which usually operate automatically but may be overridden temporarily by the higher (conscious) centers of the brain. Sensors known as *chemoreceptors* continually deliver information to the respiratory centers about how much oxygen and carbon dioxide the arterial blood is carrying. Other sensors known as *mechanoreceptors* inform the respiratory centers about the state of inflation of the lungs. The information supplied by these various sensors is used by the respiratory centers to control the actions of the respiratory muscles in order to produce a level of ventilation appropriate to the body's needs. The neural control of respiration is thus based on negative feedback, and is normally able to maintain the partial pressures of arterial oxygen and carbon dioxide within very tight bounds, even during exercise.

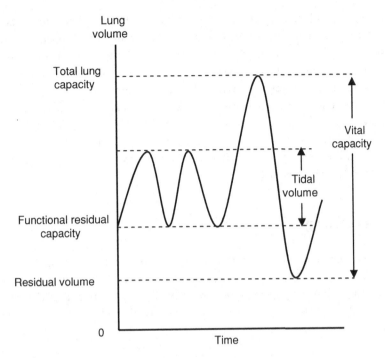

Figure 1.2 The standard subdivisions of lung volume.

1.2.3 Lung mechanics

A healthy blood-gas barrier and a functioning respiratory control system are not enough to guarantee effective gas exchange, however, unless the mechanical properties of the lung are also up to the task. These properties, basically the *flow resistance* of the airways and the *elastic recoil forces* of the lung tissues, must be successfully overcome by the respiratory muscles with each breath if fresh air is to be supplied to the blood-gas barrier. The resistance of the airway tree is determined by the internal dimensions of its various branches. The upper airways comprise the nose, mouth, and pharyngeal regions. The pulmonary airway tree starts on the other side of the vocal cords, beginning with the trachea and proceeding though a series of bifurcations to reach the *terminal bronchioles*. Each terminal bronchiole leads into an *acinus*, frequently depicted (Fig. 1.1) as something resembling a bunch of grapes (alveoli) on a set of branching twigs (the respiratory bronchioles). Exchange of gases between air and blood takes place within the acinar regions of the lung, so the acinus can be considered the basic ventilatory unit.

The conducting airways are lined with a delicate *epithelium* that partakes in numerous metabolic activities. Some of the cells in the epithelium continually secrete protective mucus that, being sticky, acts to trap inhaled particles of potentially noxious materials. The mucus and its particle prisoners are then swept up to the tracheal opening by tiny hair-like projections known as *cilia*. The cilia project into the airway lumen from specialized epithelial cells, and beat in the direction of the tracheal opening. The walls of the airways also contain significant amounts of *smooth muscle*. In the trachea this smooth muscle exists as a continuous band along its posterior aspect, linking the open ends of cartilage horseshoes that give the trachea its mechanical stability. Contraction of the tracheal smooth muscle causes the open ends of the horseshoes to approach each other, thereby decreasing the cross-sectional area of the tracheal lumen. Smooth muscle is wrapped more or less circumferentially around the airways distal to the trachea, and extends as far down as the alveolar ducts that serve as the entrance to the gas-exchanging zone of the lung. Whether or not there is any survival advantage to having smooth muscle in our lungs is still debated, but a disease such as asthma leaves little doubt that its presence can have adverse consequences.

1.3   Pathophysiology

There are numerous diseases that can make it difficult for the lungs to do their job, one way or another. Our focus is on pathologies that involve mechanical abnormalities. These pathologies constitute a substantial fraction of the public health burden in all age groups in modern society, and are classically divided into two categories labeled *obstructive* and *restrictive*.

1.3.1   Obstructive lung disease

The archetypical example of an obstructive lung disease is *asthma*, a common syndrome that varies widely in severity. Asthma is defined on the basis of its functional characteristics. Principal among these is reversible airway obstruction, demonstrated as an improvement in lung function following treatment with drugs that relax airway smooth muscle. The definition of asthma also requires a degree of inflammatory involvement as indicated by the presence of certain types of cells in the airway secretions that are brought up by coughing. However, the mechanisms underlying the pathophysiology of asthma are still debated. Indeed, it seems very likely that asthma represents the common clinical endpoint for a number of different pathological processes. Nevertheless, persistent inflammation in the lung is involved in many cases of asthma, and the inopportune contraction of airway smooth muscle is clearly a key event in an acute asthmatic attack. As most people probably recognize, the chief characteristics of asthma are wheezing and shortness of breath. Curiously, in recent times, the incidence of asthma has increased markedly in Western nations for reasons that remain poorly understood, although prevalence seems to have leveled off over the last decade.

Another common obstructive pathology is the condition known as *chronic obstructive pulmonary disease* (COPD), which frequently follows from a lifetime of heavy smoking. This, again, is a complex disease exhibiting a spectrum of features. Prominent among these is *emphysema*, which involves the progressive destruction of the microstructure of the lung tissue. The result is a reduction in the surface area of the blood-gas barrier that in mild cases may simply limit exercise capacity, but when severe may confine a patient to complete inactivity and a dependency on supplemental oxygen. The main mechanical consequence of emphysema is a reduction in the elastic recoil of the lung tissue.

The pulmonary airways are not completely rigid. This makes them prone to collapse during vigorous expiration, even in normal individuals who exhale forcefully enough. Indeed, healthy individuals with normal abdominal and thoracic muscle strength are able, over most of the lung volume range, to exhale at a rate that cannot be exceeded despite increases in expiratory effort. This maximum exhalation rate can be determined by measuring the flow of gas leaving the mouth. The magnitude of the maximum flow during a forced expiration is reduced in both asthma and emphysema. In extreme cases, the limiting flow may even be attained during quiet breathing.

The phenomenon of *expiratory flow limitation* is exploited in the diagnosis of obstructive lung diseases because the maximum volume of gas that can be expelled from the lungs during the first second following a maximal inspiration, known as forced expiratory volume in one second (FEV_1), is reduced in these diseases. On the other hand, the total volume forcibly expelled over the course of an entire expiration, the so-called *forced vital capacity* (FVC), remains relatively unaffected in a purely obstructive disease. In other words, all the air comes out eventually, but it just takes longer than normal to do so. A graphic representation of flow limitation is provided by plotting flow against expired volume throughout the entire course of a maximal expiratory maneuver. In obstructive lung disease, the expiratory *flow-volume curve* is depressed and is frequently concave upwards compared to the normal, relatively straight curve (Fig. 1.3).

1.3.2   Restrictive lung disease

The other major class of pathologies affecting lung mechanics, the so-called restrictive diseases, is exemplified by pulmonary fibrosis. Here, aberrant deposition and organization of connective proteins, particularly collagen, leaves the lungs scarred and stiff with a reduced capacity to accommodate inspired air. A reduced inspiratory capacity is also found in situations where the *surface tension* of the liquid that lines the airways and alveoli is increased. Normally, this surface tension is maintained at low levels by the presence of *pulmonary surfactant*, a detergent-like molecule secreted by cells in the lining of the airways and alveoli. The efficacy of surfactant can be reduced by leakage of plasma fluid and proteins from the pulmonary blood vessels into the airspaces of the lung, as can occur in pneumonia or pulmonary edema. Although not usually considered a restrictive condition, fluid accumulation in the airspaces can nevertheless flood some lung regions completely, effectively shutting them down. Such events decrease the total air volume of the lung and increase its overall stiffness, causing a commensurate reduction in the organ's capacity to inspire air.

Classically, restrictive lung diseases are said to be typified by a reduction in the amount of gas that can be drawn into the lungs during a maximal inspiratory effort, while the shape of the maximum expiratory flow remains relatively normal. This produces an expiratory flow-volume curve that intersects with the normal curve over a truncated volume range (Fig. 1.3). The simple view of things is thus that obstructive and restrictive lung diseases are separable on the basis of the kinds of expiratory flow-volume curves they produce. In reality, things are not quite this simple; many lung pathologies with mechanical manifestations exhibit varying degrees of both obstructive and restrictive

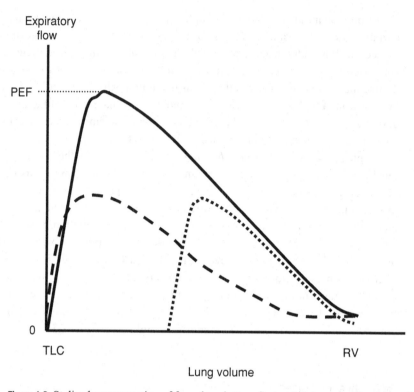

Figure 1.3 Stylized representation of forced expiratory flow-volume curves for a normal lung (solid line), an obstructed lung (dashed line), and a restricted lung (dotted line). The normal peak expiratory flow (PEF) is reduced in both pathologies, but only in the restrictive case is total lung capacity (TLC) commensurately reduced. This highly simplified diagram shows residual volume (RV) being identical in all cases, but RV is frequently increased in obstructive disease due to air becoming trapped behind airways that close as lung volume decreases.

patterns. What is certain, however, is that abnormalities in lung mechanical function accompany a wide and important range of pathologies.

1.4   How do we assess lung mechanical function?

The assessment of respiratory mechanics involves uncovering relationships between key pressures, flows, and volumes measured at appropriate sites. Accordingly, what we know about lung mechanical function is dictated by what we can measure. In this regard, the lung has long been a favorite organ of study for quantitative scientists, including mathematicians and biomedical engineers, because it is relatively easy to obtain data from it. For example, gas pressures, flows, and volumes at the mouth are readily monitored with high accuracy and temporal resolution. Controlled perturbations in these variables can be easily applied as probes to investigate the lung's internal workings. Nevertheless, most events taking place inside the lung that influence its mechanical function are not accessible by direct observation. This leaves us having to infer what is going on from dynamic relationships observed between those limited variables that can be measured experimentally. Ideally, when we observe abnormal relationships between these variables, we would like to be able to deduce what structural abnormalities caused them.

Some inferences about the internal workings of the lung can be made by observing pressures, flows, and volumes and using a little common sense. For example, a sharp increase in peak airway pressure during mechanical ventilation likely indicates that something has suddenly impeded the flow of air into the lungs. Even though an elevated peak pressure on its own gives no information as to where the airway impediment might be, it may still prompt the anesthesiologist to check the endotracheal tube and find that it is blocked with mucus – clearly a useful outcome. Empirical quantities such as peak airway pressure, FEV_1, and FVC that are derived from measured variables may thus have great utility despite being of limited specificity.

The full inferential potential hidden inside measurements of respiratory pressures, flows, and volumes, however, is only revealed once we use them to derive quantities based on theoretical models. Perhaps the most immediate example of this is the calculation of pulmonary *airway resistance* as the ratio of the pressure drop between the proximal and distal ends of the airway tree to the air flow through it. Leaving aside for the moment the question of how one actually measures these pressures and flows, it is obvious that the calculation of airway resistance is motivated by the notion that the airway tree behaves like a single rigid conduit through which flows an incompressible fluid. This is a very simple model of what is in reality a very complicated structure. Nevertheless, this model has proven enormously useful because it allows a measure of function (resistance) to be linked to hypothetical structure (airway length and internal diameter). The concept of airway resistance is also readily accessible to the educated mind without the need for special analytical tools. Even so, it is obvious that one ought to do much better by using a model with a structure more closely resembling the anatomy of a real airway tree. As the complexity of such a model increases, however, predicting the details of its behavior rapidly exceeds the capacity of the unaided human intellect.

To break free of the constraints of human intuition in developing a quantitative understanding of lung mechanics, we must resort to the systematic construction of mathematical models. A mathematical model is a set of equations that serve both as a precise statement of our assumptions about how the lung works mechanically and as a means of exploring the consequences of those assumptions. As alluded to above, the human mind is incapable of doing either without the aid of mathematical tools except in the most trivial of cases. A state-of-the-art understanding of lung mechanics thus requires a certain familiarity with the methods of mathematical and computer modeling. This requires some effort, but is well worth the reward of enlightenment that ensues.

1.4.1   Inverse modeling

The process of trying to construct a mathematical model of a system from measurements of *inputs* to and *outputs* from the system is known as *inverse modeling* (also known as *system identification*). The *parameters* of the model are evaluated by getting the model to predict the outputs from the inputs as accurately as it can. The structure of such a model is not generally known *a priori*, so it has to be determined from considerations of experimental data, prior knowledge about the structure of the system being modeled, and whatever else can be brought to bear on the issue. The modeler uses all this information to specify what the various components of the model are, and how they are to be linked together. Ideally, the structure of an inverse model should correspond in some useful way to the structure of the real system, so that when the model mimics the behavior of the system it does so in a way that is true to the internal mechanisms responsible for the behavior of the system itself.

For obvious practical reasons it is usually not possible to include every known component of a complicated system in a mathematical model. This is certainly true in the case of the lung, even if we understood in detail how each of its components works (which we don't). Therefore, the choice of model structure requires a decision about which components of the real system are important for the purpose at hand, and which components can be safely ignored. This is not a process which is readily codified. Indeed, determination of model structure is very much an art that reflects the experience and wisdom of the modeler. It is also a dynamic process; models of complex systems such as the lung are constantly being tested and refined in the light of new experimental data and knowledge.

Mathematical models do not have to be complicated to be useful. Indeed, inverse models are invariably rather simple, having few independent components and small numbers of adjustable parameters. This is a necessary consequence of the fact that such models have to be matched to experimental data, and there are usually only so many free parameters that even the most precise data can support. Inverse models of the lung, for example, do not even come close to encapsulating everything we know about the organ, yet they are still capable of mimicking many of the details of its global behavior. An example of an extremely simplistic but nevertheless intuitively acceptable model of lung mechanics consists of an elastic balloon sealed over the end of a rigid pipe; the balloon represents the expandable lung tissues while the pipe represents the pulmonary airways (Fig. 1.4). Obviously, a real lung is vastly more complicated than this simple construct, even though it still embodies much that is key to the process of ventilation.

Once the structure of an inverse model of the lung has been settled upon, the mathematical equations describing its mechanical behavior – the so-called *equations of motion* – must be derived. These equations state how pressure is related to flow and volume within each component of the model, and tell us exactly how the complete model will behave under every conceivable circumstance. The world of mathematical models is thus fundamentally different from the real world in which we breathe. In the real world we can never measure anything exactly, nor understand any system down to the last detail. By contrast, in the world of models it is possible to know everything there is to know about a particular model.

Equations of motion contain quantities known as *variables*. These represent the things that are measurable, and which usually vary with time. The variables in models of lung mechanics are typically gas pressures, flows, and volumes. Equations of motion also

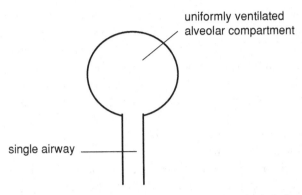

uniformly ventilated
alveolar compartment

single airway

Figure 1.4 The lung is modeled most simply as an elastic balloon at the end of a rigid pipe. The balloon represents the distensible tissues while the pipe represents the conducting airways that connect the mouth and nose to the alveolar regions of the lung.

contain things known as *parameters* that have fixed values and that characterize the physical attributes of the parts of the real system that the various model components represent. Typical parameters one might find in a model of lung mechanics include the diameter or flow resistance of an airway, or the elastic stiffness of a piece of lung tissue. The behavior of a model is tuned by adjusting the values of its parameters. The process of finding those parameter values that cause the model to behave like a particular real lung is known as *parameter estimation*.

An inverse model must always be viewed as a work in progress. No matter how successful a particular model might be, it will always have shortcomings. This goes without saying because we can never hope to perfectly delineate the quantitative dynamic behavior of something as complicated as a lung. The best we can hope for is that an inverse model is adequate for a particular purpose. The process of establishing that this is the case is known as *model validation*, and involves comparing the predictions of the model to appropriate experimental data. If the model predictions match experimental observation well enough, then the model may be judged acceptable. If, on the other hand, the differences between the model and system behaviors are too big to ignore, then the model must be discarded and a different (and invariably more complicated!) model used in its place. Acceptance or rejection of a model is often based on statistical criteria, but these involve arbitrary decision thresholds and assumptions that are frequently poorly met. In the final analysis, model validation invariably comes down to a judgement call.

1.4.2 Forward modeling

The counterpart of inverse modeling is known as *forward modeling* (also known as *simulation*). This involves calculating the output of a system from a given input on the basis of a model of the system. This is fundamentally different from inverse modeling because here we are not trying to figure out what is going on inside the "black box." Instead, we have already decided what is inside the box, and we are using that knowledge to predict behavior. As explained above, inverse models can only ever reach a certain (usually rather low) level of complexity before their components become too numerous to interpret unambiguously and their parameters too numerous to estimate reliably from the available experimental data. Forward models suffer no such constraint and in principle can include as much detail as is available in the form of *a priori* knowledge about the system being modeled. It is possible, for example, to make a forward model of the lung that includes the flow resistance of every single branch in the airway tree, which in a human lung number in the tens of thousands. By contrast, when fitting an inverse model of the lung to experimental data, it is usually difficult to uniquely identify the resistances of more than two or three distinct airways.

Because of these differences, forward models can serve as powerful tools for the development of inverse models by allowing specific hypotheses to be tested in a way that is often impossible experimentally. Suppose, for example, that the application of an aerosolized drug to the lungs of a mechanically ventilated subject causes changes in the relationship between gas pressure and flow measured at the mouth. Further suppose that the aerosol particles are of a size known to be deposited far down in the lung. One might suspect on the basis of this information that only the most distal airways would have constricted in response to the drug. This is probably impossible to verify directly with current technology. With an anatomically accurate forward model of the lung, however, one can simulate pressure-flow relationships when only the distal airways are narrowed in order to see if this reproduces the relationships that were measured experimentally. Of course, such an exercise will never prove the hypothesis right or wrong because the forward model can never mimic the real lung perfectly. Nevertheless, a good forward model can lend significant support to the acceptance or rejection of a hypothesis of this nature. This establishes consistency between the hypothesis and the state-of-the-art knowledge about the lung that is embodied in the forward model.

1.4.3   The modeling hierarchy

The study of lung mechanics is about linking structure to function; experimental measurements of mechanical function are made in the laboratory and then used to infer something about the structure of the lung itself. Our understanding of the structure-function link is encapsulated in models – idealizations that we believe embody the important aspects of the lung, and that we can wrap our minds around. The tools of mathematical modeling allow us to take this process to levels of complexity and precision far beyond that achievable by human intuition alone. Mathematical inverse models that are identified directly from experimental data, and computational forward models that are constructed from prior knowledge about lung structure, together constitute the totality of our understanding of lung mechanical function. The science of lung mechanics thus progresses through the continual interplay between the two modeling paradigms (Fig. 1.5).

As a result of the progressive nature of model development, inverse models of the lung can be arranged in a hierarchy of complexity as illustrated in Fig. 1.6. Exploring this hierarchy is the central theme of the rest of this book, and begins with the simple model shown in Fig. 1.4. It is difficult to imagine a simpler structure than this that could

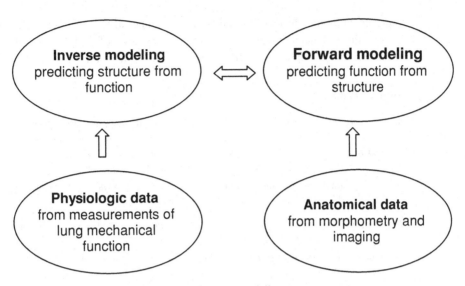

Figure 1.5 Synergy between forward and inverse modeling.

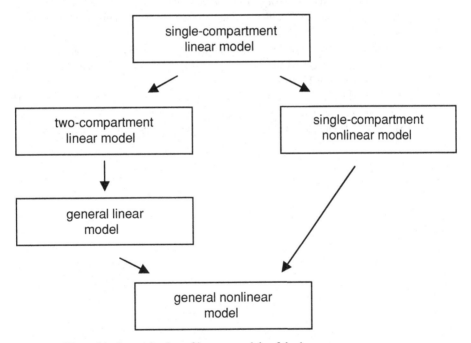

Figure 1.6 Hierarchical organization of inverse models of the lung.

usefully represent lung mechanics in some kind of overall fashion. How to proceed to the next stage of complexity, however, is not immediately obvious. One possibility is to require the elements of the simple model to be nonlinear. Alternatively, we could add a second linear compartment, so taking even the first step in model sophistication presents a dichotomy. Above two-compartment models lies a third level of complexity represented by models that have an arbitrary number of compartments, comprising the general linear dynamic model. The fourth and final level of complexity is attained when we allow both multiple compartments and nonlinear behavior.

Further reading

The very brief overview of lung anatomy, physiology, and pathophysiology given above is merely to set the scene and establish context for the remainder of the book. It is by no means an exhaustive account of any aspect of these complicated topics, the details of which are covered in detail in numerous excellent texts. The following are some notable examples.

Highly readable and succinct accounts of basic pulmonary physiology and pathophysiology are contained in the following two books by John B. West:

Respiratory Physiology. The Essentials. 8th edition. Philadelphia: Lippincott Williams & Wilkins, 2007.
Respiratory Pathophysiology. The Essentials. 7th edition. Philadelphia: Lippincott Williams & Wilkins, 2008.

Although now 20 years old, an in-depth treatment of all aspects of pulmonary physiology can still be found in the *Handbook of Physiology* published by the American Physiological Society, Bethesda, Maryland. Of particular relevance are the two volumes devoted to the mechanics of breathing (Section 3, Volume III, parts 1 and 2).

A more modern but still detailed treatment is to be found in *Physiologic Basis of Respiratory Disease*, edited by Qutayba Hamid, Joanne Shannon, and James Martin, published by BC Decker, Hamilton, Ontario, 2005.

For an exhaustive treatment of pulmonary pathophysiology and medicine, the reader can consult *Textbook of Respiratory Medicine*, 3rd edition, edited by John F. Murray, Jay A. Nadel, Robert J. Mason, and Homer A. Boushey, Jr., published by W. B. Saunders, Philadelphia, 2000.

2 Collecting data

2.1 Measurement theory

The usefulness and validity of any model of lung mechanics rest entirely on the experimental data upon which it is based. An appreciation for these data and an understanding of how they were obtained are therefore vital for a complete understanding of the model itself.

The experimental data required for the construction of models of lung mechanics usually consist of gas pressures, flows, and volumes. These *variables* can be measured at a variety of sites around the body. By far the most common measurement site is at the entrance to the airways (this is the nose and mouth in an intact subject, but may be the entrance to the trachea in experimental animals or patients receiving mechanical ventilation). However, other sites have been used, such as at the body surface, inside the esophagus, and even within individual alveoli. Generally speaking, increasing the number of simultaneous measurement sites allows for an increase in the complexity of the possible models that can be identified from the resulting data. Of course, this has to be balanced against practical and ethical considerations.

The measurement of a variable such as pressure occurs in a sequence of steps, as depicted in Fig. 2.1, beginning with the variable itself and ending with the recorded data. First, the pressure is allowed to impinge on a pressure *transducer*, which is a device that converts the pressure signal into a corresponding voltage signal. This is invariably accompanied by the addition of some unwanted energy, termed *noise*, arising from any uncontrolled aspect of the experiment. The noisy voltage signal is then usually *conditioned* in some fashion to get it ready to be recorded. Signal conditioning generally consists of *amplification* and *filtering*. The signal is then *sampled* at regular time intervals by an *analog-digital (A-D) converter*. This is an electronic device that converts a voltage level into an integer number. The resulting string of integers is then stored on a computer until it can be used in the process of constructing a mathematical model of the lung.

2.1.1 Characteristics of transducers

There is no such thing as a perfect transducer, one that produces a perfectly faithful voltage representation of some variable of interest. Real transducers vary substantially, however, in their degrees of imperfection. In some cases, a transducer may be of such high quality that its imperfections can be ignored with no consequence to the scientific

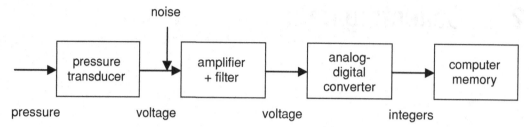

noise

| pressure transducer | amplifier + filter | analog-digital converter | computer memory |

pressure   voltage   voltage   integers

Figure 2.1 The general measurement scenario in which a pressure signal is converted into a set of digital data stored on a computer.

conclusions that are drawn from its use. In other cases, the voltage output from a transducer can be a severely distorted representation of the input variable that it purports to measure, requiring either that the output be corrected in some fashion or the transducer be discarded. It is therefore crucial, in any experimental situation, to understand the imperfections of the transducers being used to collect data.

There are various *static properties* that characterize a transducer's performance. These properties pertain when the measured signal changes slowly relative to the transducer's ability to keep up with it. The most obvious of these properties is *linearity*, which refers to the extent to which the voltage v produced by a transducer is proportional to the biological signal (again we will use pressure, P, as our example). If the transducer is linear then v and P are related by an equation of the form

$$v = aP + b \tag{2.1}$$

where a is the constant of proportionality and b is a constant defining the offset (Fig. 2.2A). Linearity was a highly desirable trait in the days before widespread availability of digital laboratory computers because it meant that P could be determined from v by inverting Eq. 2.1, an essentially trivial manual operation. With a computer, virtually any nonlinear relationship between v and P can be inverted instantly provided the relationship is single-valued (such as the dashed line in Fig. 2.2A). Linearity of transducers operating under quasi-static conditions has thus become less of an issue.

Another important static property that a transducer may possess is *hysteresis*. This troublesome property arises when the value of v corresponding to a particular value of P depends on whether P was approached from above or below (Fig. 2.2B). In contrast to nonlinearity, hysteresis is usually extremely difficult to correct for even with a computer, so one should always try to use a transducer with minimal hysteresis.

In selecting a transducer for a particular application, it is also important to make sure it has the appropriate *resolution* and *dynamic range*. These characteristics determine the smallest change in P that the transducer can detect, as well as the largest change it can record without saturating. The *signal-to-noise ratio* is another important transducer characteristic that influences its resolution, and is defined as the ratio of the magnitude of the desired voltage signal to the magnitude of the noise (Fig. 2.1). Noise is unpredictable, so its magnitude must be substantially lower than any of the changes in the signal that need to be measured (ideally by at least an order of magnitude). If the signal-to-noise

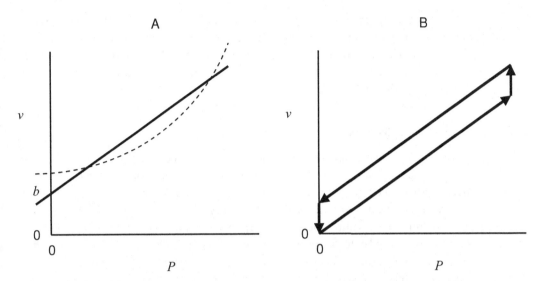

Figure 2.2 (A) A linear transducer follows the solid line relationship between P and v, while a nonlinear one might follow the dashed line. (B) Hysteresis occurs when the value of v depends on the direction from which a particular value of P is approached.

ratio is not sufficiently high, it will not be possible to tell if an observed change in v actually reflects a corresponding change in P.

Transducers also have *dynamic characteristics* that determine how well they perform when recording signals that vary with time. The ability of a transducer to follow a time-varying signal is encapsulated in its *frequency response*, a complete understanding of which requires the tools of linear systems theory and the Fourier transform. We will meet these tools in Chapter 8 because it turns out that they are also required to analyze the dynamic mechanical behavior of the lung. For the time being, however, suffice it to say that a transducer's frequency response defines how well it manages to produce a v mirroring P when P is oscillating at a particular frequency. To put this in quantitative terms, suppose P is a single sine wave with unity amplitude oscillating at frequency f Hz,

$$P(t) = \sin(2\pi f t). \tag{2.2}$$

Ideally, we would like v to be the same (following appropriate calibration). In general, however, we find that

$$v(t) = A \sin(2\pi f t + \phi). \tag{2.3}$$

In other words, v is still a sine wave oscillating at frequency f Hz, but its amplitude has been scaled by some factor A and its phase has been shifted by an amount ϕ. Provided the transducer is linear, A and ϕ do not depend on the amplitude of P, but they may vary markedly with f. Indeed, the way that A and ϕ vary with f constitutes the frequency response of the transducer, and leads to it being labeled as either *low-pass* or *high-pass*. Pressure transducers, for example, are invariably *low-pass* systems because they can respond faithfully to P when it varies slowly, but they have increasing difficulty keeping up as the variations in P increase in frequency. In other words, A is close to 1 when f is small, but it decreases toward zero once f exceeds a certain level. *High-pass* transducers operate in reverse fashion; A approaches zero as f decreases below a certain threshold, but above the threshold A is close to 1.

2.1.2   Digital data acquisition

Once a transducer has produced its voltage output, v, the data must be recorded and stored for subsequent analysis (Fig. 2.1). Before the days of laboratory digital computers, one went straight from v via an amplifier to final recording. In the first half of the twentieth century, this was achieved with a *smoked-drum kymograph*. This device used v to control the vertical position of a sharp metal pointer that scratched a visible line through soot coated onto the surface of a rotating cylindrical drum. This was superseded by the *electronic chart-recorder* in which the metal pointer was replaced by one or more ink pens, and the smoked drum was replaced by a long sheet of paper scrolling by at a constant speed. The primary data record produced by a chart-recorder typically consisted of a (frequently large) length of fan-folded paper containing tracings corresponding to the various time-varying signals measured in the experiment. This may seem quaint and archaic by today's electronic data standards, but the chart-recorders in wide use in physiology labs until quite recently were sophisticated and precise machines capable of a high degree of recording fidelity. They also provided the experimenter with an immediate hardcopy graphical representation of the data that could be conveniently annotated as the experiment progressed. Even so, the resolution and dynamic range of a chart-recorder is nowhere near that of even the most modest computerized data acquisition system available today. Also, quantitative analysis of data plotted on a paper record has to be performed manually, which obviously places a severe practical limit on what can be done. Over the past few decades, the digital computer has come to completely replace these earlier analog recording devices, and consequently to revolutionize the way that physiological research is done.

The next step in the data acquisition process after the transducer has done its job is thus to *digitize* the analog voltage signal. This is achieved with an A-D converter, which essentially determines the instantaneous value of v at regularly spaced time intervals. The continuous voltage signal is thus converted into a string of numbers. These numbers and their locations in time constitute the *discretized* version of the original analog signal.

There are several important factors to consider when using an A-D converter. As with transducers, resolution and dynamic range are paramount because they determine the smallest difference in v that the A-D converter can distinguish, as well as the maximum variation in v that can be faithfully recorded. The allowable voltage range that v can occupy without saturating the A-D converter is divided into equally spaced bins numbered from 1 to 2^N, where N is the number of *bits* in the A-D converter. A 12-bit A-D converter has $2^{12} = 4096$ bins, so if its range is 0–10 volts then it can record voltage differences of $10/4096 = 0.0024$ volts. Any value of v between 0 and 0.0024 volts is

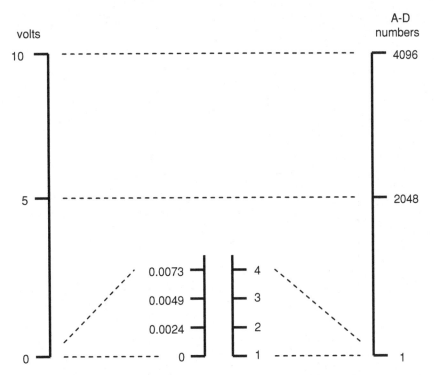

Figure 2.3 A 12-bit A-D converter with an input voltage range of 0–10 volts converts a signal within this range into a string of numbers having values between 1 and 4096.

assigned to bin number 1, any value between 0.0024 and 0.0049 volts is assigned to bin number 2, and so on (Fig. 2.3).

Suppose, for example, one wants to measure the pressure, P, applied to a patient's lungs, and that P may vary between 0 and 30 cmH$_2$O. (This is the unit of pressure favored historically by the lung mechanics community; 1 cmH$_2$O is the pressure exerted by the weight of a column of water 1 cm in height. The cmH$_2$O is a convenient unit because it is easy to assemble a pressure calibration device, called a *manometer*, consisting of a vertical tube containing a column of water, the height of which is readily adjusted. The cmH$_2$O continues to be widely used despite its archaic heritage. Fortunately, 10 cmH$_2$O is very close to one kilopascal, an SI unit of pressure.) If P is to be recorded on a 12-bit A-D converter, the resulting digitized signal will have a maximum resolution of $30/4096 = 0.007$ cmH$_2$O, which probably exceeds the resolution one might require for any conceivable study of lung mechanics. The same applies to the great majority of pulmonary applications involving the measurement of flow and volume.

These favorable circumstances concerning resolution only pertain, however, when a significant fraction of the voltage range of the A-D converter is used. A common error in the laboratory occurs when the voltage signal produced by a transducer is not amplified sufficiently, so that only a few of the bins in the A-D converter are utilized. For example, if the A-D converter has an input range between 0 and 10 volts, but v is confined between

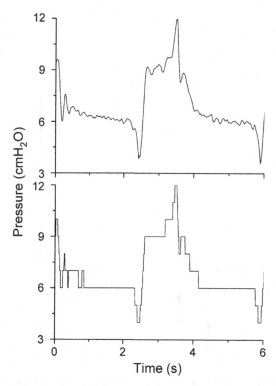

Figure 2.4 The top panel shows a high-fidelity recording of pressure obtained at the entrance to the endotracheal tube in a patient receiving mechanical ventilation. The lower panel shows what the recording could look like if the analog voltage signal coming from the pressure transducer was not amplified sufficiently to fully utilize the dynamic range of the A-D converter.

0 and 1 volt, then the effective resolution of the A-D converter has been reduced tenfold. In the extreme case where the total excursion in v corresponds to only a handful of adjacent bins in the A-D converter, the digitized signal will be seen to jump vertically between discrete levels as illustrated in Fig. 2.4. When this happens, the digitized signal contains *discretization errors* and consequently is no longer a faithful representation of the original signal.

2.1.3   The sampling theorem and aliasing

We have just considered how the vertical resolution of an A-D converter impacts the recording of a biological signal such as pressure. But what about the temporal resolution? The pressure applied to the lungs of a mechanically ventilated patient is a continuous signal. As such, it is composed of an infinity of points over any finite time interval, yet even the fastest A-D converter can collect only a finite number of data points. Does this mean we always face loss of information when digitizing a continuous signal? Fortunately, the answer to this question is no, thanks to something called the *sampling theorem*, which can be understood as follows.

Most continuous signals tend to be quite irregular (e.g. Fig. 2.4), so they do not look much like a sine wave with its regularly repeating undulations. However, the irregularities in any signal can be viewed over a range of different time scales; viewed up close the variations in a signal may appear as rapid oscillations, while when you take step back and look at the overall trends they often appear as slower undulations on a larger scale. In other words, an arbitrary signal can be viewed as having a number of different components, some fast and some slow, that add together to produce the signal itself. This being the case, it should be possible to approximate any analog signal to some degree of accuracy as a sum of sine waves, each of the form of Eq. 2.3. That is,

$$P(t) \approx \sum_{i=0}^{N} A_i \sin(2\pi f_i t + \phi_i). \qquad (2.4)$$

It turns out that this description becomes exact as the spacing between adjacent frequencies in the sum shrinks toward zero, requiring in turn that N in Eq. 2.4 tends to infinity. The general continuous signal can thus be expressed as the integral of a continuous distribution of sine waves, thus:

$$P(t) = \int_{0}^{f_0} A(f) \sin(2\pi f t + \phi(f)) df \qquad (2.5)$$

where A and ϕ are now functions of f. $A(f)$ quantifies how much each sine wave contributes to $P(t)$, and therefore defines its *frequency content*.

For some signals, the upper limit of the integral, f_0, in Eq. 2.5 is effectively infinite. For other signals, however, f_0 is finite (and there is still an infinity of values of f between it and zero). When f_0 is finite, the signal is said to be *band-limited*. In this case, the sampling theorem says that all the dynamic information the signal contains is faithfully captured by sampling it at a frequency greater than twice f_0. This means that the number of data points has to be at least as great as the number of times the signal reverses vertical direction (recall that, from Eq. 2.5, the most rapid oscillations in the signal come from the sine wave component at f_0). Thus, by making the data sampling frequency greater than $2f_0$, known as the *Nyquist rate*, we lose no information. The entire original continuous signal can be reconstructed from the sampled points alone provided they are sampled at or above the Nyquist rate.

A practical issue arises with respect to the number of data points that need to be collected in order to satisfy the sampling theorem. Obviously, this number can become unmanageably large if f_0 itself is too large. However, the temptation to *under-sample* the signal must be resisted at all costs because dropping the sampling frequency below $2f_0$ does not simply sacrifice the information contained in the higher frequencies. Instead, this information reappears at lower frequencies in the sampled data. This phenomenon is known as *aliasing*, and is illustrated in Fig. 2.5.

Aliasing is particularly dangerous when the nature of the spectral content of a signal is central to its interpretation. This is the case, for example, when respiratory pressures and flows are recorded for the purpose of calculating the mechanical impedance of the lungs (described in Chapter 8). Because one can never guarantee that these signals will

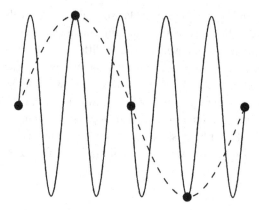

Figure 2.5 A continuous sine wave (solid line) oscillating with a frequency f_0 is sampled (dots) with a frequency less than $2f_0$, yielding a set of discrete points that define an aliased sine wave (dashed line) that has a frequency much less than f_0.

be band-limited below some manageable f_0, it is always necessary to pass them through high-quality, low-pass, electronic *anti-aliasing filters* before they are digitized. This enforces a desired value of f_0. It is crucial to remember, though, that filtering must be done *prior* to digitization; once the signals have been sampled, any aliasing that occurs cannot be undone.

2.2   Measuring pressure, flow, and volume

Having established the general considerations for measuring signals in the laboratory, we now consider how one measures those signals that are specific to the study of lung mechanics.

2.2.1   Pressure transducers

Pressure is always measured in terms of the mechanical deformation it causes in some elastic material. Pressure transducers differ merely in the particulars of the deformable element and how its deformation is recorded. The transducers used for measuring gas pressures in respiratory applications come in two basic configurations: gauge and differential. A gauge transducer references the pressure of interest to atmospheric pressure, while a differential transducer compares two test pressures (Fig. 2.6).

Until relatively recently, the mainstay of pressure measurement in the respiratory physiology laboratory was the *variable reluctance transducer*. This type of transducer operates like an AC transformer. The primary coil of the transformer is excited by several kHz of alternating electric current that then induces a voltage in the secondary coil. The efficiency of this induction is influenced by the configuration of a thin metal disk placed between the two coils. A pressure applied to one side of the disk causes it to deform, thus altering the induced voltage in the secondary coil. The magnitude of the induced voltage is then read out as a signal proportional to the pressure. Variable reluctance

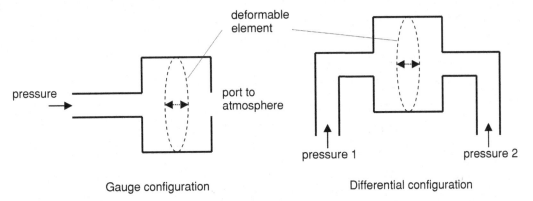

Gauge configuration   Differential configuration

Figure 2.6 Gauge and differential pressure transducers. The application of a pressure difference deforms an elastic element. The degree of deformation is then converted to a voltage.

transducers are accurate and can be imbued with varying sensitivities depending on the elastic stiffness of the metal disk. They have frequency responses that are flat out to 20 Hz or so, depending on the length of the tubing used to connect them to the site of pressure measurement [1]. This is reasonably good relative to the frequency content of typical respiratory pressure signals. However, these transducers are somewhat cumbersome and can be damaged by over-pressurization. Also, the electronic circuits required to excite them and demodulate their AC responses to pressure are somewhat involved.

While the variable reluctance transducer has a venerable history in lung mechanics research, it has now largely been replaced by the *piezoresistive* transducer [2]. Here, the pressure sensitive element is a tiny strain gauge that changes its electrical resistance in proportion to its deformation. A constant voltage (or current) is passed through the piezoresistive element, which is configured to comprise one of the four arms of a suitably balanced *Wheatstone bridge*. This is a very simple circuit composed of four resistors across which voltage is a linear function of one of the resistors, in this case the piezoresistive element, when the values of the other three resistors are held fixed. The pressure impinging on the deformable element thus becomes proportional to the voltage across the bridge. All that remains at this point is to amplify this voltage and pass it through an anti-aliasing filter prior to digitization. Piezoresistive pressure transducers first started to become widely used in respiratory research in the 1980s. Initially, they tended to suffer from baseline drift, were affected by changes in orientation and temperature, and had limited sensitivity. These problems have now been overcome, allowing piezoresistive transducers to be exploited for their several advantages. These include an extremely high frequency response (typically flat to several hundred Hz), robustness (they can be pressurized to many times their nominal full-scale range without damage), and the fact that they can be manufactured using solid-state technology to be very small, light, and cheap.

2.2.2 Measuring lateral pressure

The assessment of lung mechanics generally requires that pressure be measured at some point in a flowing stream of gas. The goal is usually to determine the pressure that is

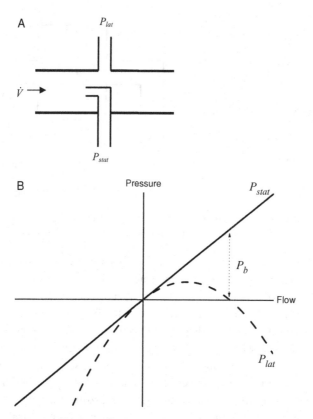

Figure 2.7 (A) Lateral pressure (P_{lat}) measured through a lateral tap, and static pressure (P_{stat}) measured with a Pitot tube. (B) P_{stat} increases linearly with flow. P_{lat} increases with flow initially, but then starts to decrease and eventually becomes negative.

driving flow at the site of measurement, because it is this pressure that is having to overcome the downstream resistance of the conduit in which the gas is flowing. It is convenient to tap laterally into the conduit and insert a pressure transducer into the tap (Fig. 2.7A). This provides what is known as *lateral pressure* (P_{lat}), which is that pressure acting perpendicularly to the direction of flow. However, P_{lat} is less than the pressure actually driving the gas along the conduit because of a phenomenon known as the *Bernoulli effect*. This phenomenon arises as a result of the principle of conservation of energy; the faster the gas moves along the tube, the larger its kinetic energy, so the less potential energy it has available in the form of P_{lat}. When the conduit has an area A and carries a flow of $\dot{V}$, then P_{lat} underestimates the driving pressure by an amount P_b, given by

$$P_b = \frac{\beta \rho \dot{V}^2}{2A^2} \tag{2.6}$$

where ρ is the density of the gas and β is a factor that depends on the flow velocity profile [3, 4]. When the profile is flat (i.e. the linear velocity of the gas molecules is the same at every point in the tube cross-section) then $\beta = 1$. When the profile is parabolic, $\beta = 2$.

In order to measure the actual driving pressure, it is necessary to bring the flow to a standstill at the site of pressure measurement. While this obviously cannot be done for the flow across the entire cross-section of the conduit, it can be achieved within a small region using a device called a *Pitot tube* that enters the flow stream laterally and then bends until its open end faces directly into the oncoming flow (Fig. 2.7A). This causes a small parcel of gas to be brought to rest as it abuts up against the open end of the Pitot tube, which then is able to measure static pressure (P_{stat}). Thus, P_b is given by the difference between P_{stat} and P_{lat}. Consequently,

$$P_{lat} = P_{stat} - P_b. \tag{2.7}$$

The problem caused by the Bernoulli effect is apparent from Eq. 2.6 and Fig. 2.7B, which show that P_b depends on the square of $\dot{V}$ and on the inverse of the square of A. When A is large enough, P_b is negligible. However, as A decreases there comes a point at which P_b starts to become significant, and P_{lat} and P_{stat} begin to diverge significantly. As $\dot{V}$ increases beyond this point, it is even possible for P_{lat} to eventually become negative (Fig. 2.7B). Thus, it is always very important to be sure that the Bernoulli effect is not significant whenever lateral pressure is being substituted for driving pressure [5], such as when pressure is measured through a lateral tap into the tubing connecting an intubated patient to a mechanical ventilator [6].

2.2.3   Esophageal pressure

The pressure drop across the lung, from the start of the airways through to the pleural space (Fig. 1.1), is an essential quantity pertaining to the study of lung mechanics. Measuring *transpulmonary pressure* is less than trivial, however, owing to the difficulty of determining *pleural pressure*. Although pleural pressure can be measured directly, accessing the pleural space is a significantly invasive procedure that is not usually feasible in human subjects and is technically challenging even in experimental animals. Fortunately, the pressure in the esophagus (P_{es}) is a useful surrogate for pleural pressure because the esophageal lumen is subjected to essentially the same pressure swings as the pleural space, being separated from it by only the soft tissues of the esophageal wall.

Measuring P_{es} requires that the subject swallow an *esophageal balloon catheter*, perhaps not an entirely innocuous procedure but nevertheless virtually risk-free and certainly better than having the chest wall punctured. The esophageal balloon catheter consists of a thin-walled balloon a few cm in length sealed over the end of a thin plastic catheter typically about 100 cm in length (Fig. 2.8). The catheter is passed through the nose and swallowed until the balloon reaches the middle of the esophagus. A small volume of air is then injected through the catheter into the balloon so that there is free transmission of pressure between the interior of the balloon and the proximal end of the catheter, to which a gauge pressure transducer is attached. The volume of air in the balloon must be sufficient to prevent its walls from occluding the hole at the end of the

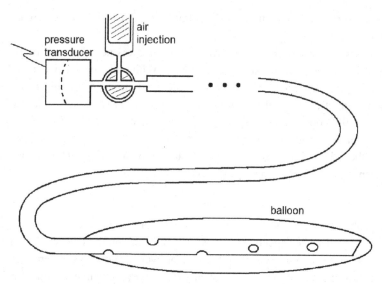

Figure 2.8 The esophageal balloon catheter. The pressure inside a latex balloon on the end of a thin catheter is sensed by a pressure transducer connected to the proximal end. A three-way stopcock permits injection of a small volume of air into the balloon so that its sides clear the multiple holes in the end of the catheter.

catheter, or any extra holes that are usually made along the segment of catheter inside the balloon. On the other hand, the balloon volume must be low enough that its walls remain flaccid, or else there will be a transmural pressure gradient across the walls.

P_{es} is not perfectly uniform along the length of the esophagus due to the gravitational gradient of the lung. Consequently, the pressure provided by the esophageal balloon tends to vary with its position in the esophagus. Nevertheless, the balloon is considered to be correctly placed when it passes the so-called *occlusion test*, which is administered by having the subject attempt to inspire against an occluded airway. Because lung volume does not change during such a maneuver, the change in pressure generated at the airway opening (P_{ao}) should be the same as that generated in the esophagus. That is, regressing P_{es} versus P_{ao} should yield a slope of unity [7]. In practice, the P_{es}–P_{ao} relationship is never perfectly linear (Fig. 2.9A), and slopes that differ from 1.0 by up to 10% are common. The occlusion test is most conveniently performed in subjects able to change P_{es} and P_{ao} through their own spontaneous breathing efforts. However, the occlusion test can also be performed by applying a uniform pressure field around the thorax; indeed, highly linear P_{es}–P_{ao} relationships have been measured using this method in paralyzed dogs [8] (Fig. 2.9B).

The frequency response of the esophageal balloon is limited by the fact that the pressure changes in the esophageal lumen must be transmitted through air along a thin catheter to a pressure transducer some distance away. However, a reasonably good response up to 30 Hz has been observed [9]. P_{es} has also been measured successfully using catheter-tip piezoresistive pressure transducers [10], which have a much better

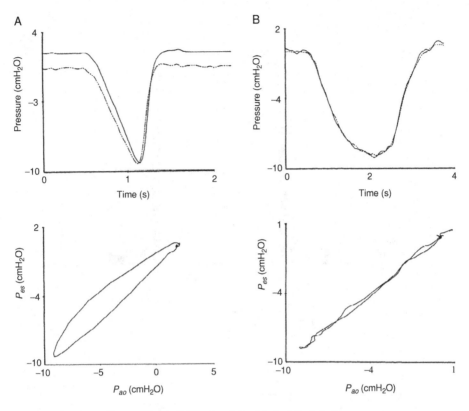

Figure 2.9 (A) Recordings of P_{ao} (solid line) and P_{es} (dashed line) during a spontaneous inspiratory maneuver in an anesthetized dog (top) and the corresponding plot of P_{es} versus P_{ao}. (B) Similar recordings from the same animal following paralysis during the application of negative pressure around the thorax (reproduced with permission from [8]).

frequency response than balloon-catheter systems. In small animals, P_{es} can be measured using a water-filled catheter [11].

2.2.4   Alveolar pressure

The pressure in the alveoli of the lung just under the pleural surface can be measured directly in open-chest animals using a technique known as the *alveolar capsule* [12]. The circular flange of a small plastic capsule is fixed to the visceral pleura (outer membrane) of the lung with cyanoacrylate glue (Fig. 2.10). The chamber of the capsule then isolates a small window on the pleural surface through which small holes are made. If the pleura is punctured carefully to a depth of 1–2 mm, preferably with a cautery needle, bleeding is minimal. This brings the sub-pleural alveoli into contact with the capsule chamber so that a small piezoresistive pressure transducer lodged in the chamber is able to give a direct recording of alveolar pressure. In an animal the size of a dog, several alveolar

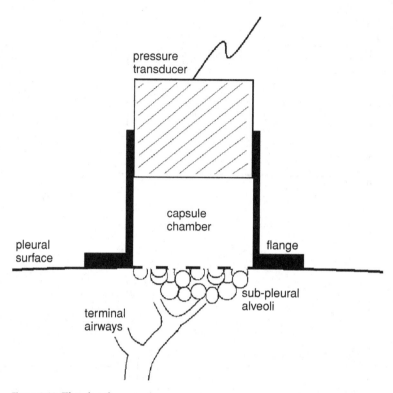

Figure 2.10 The alveolar capsule.

capsules can be installed simultaneously at different sites over the lung surface [13–15]. A single alveolar capsule can even be used in an animal as small as a mouse [16].

The alveolar capsule technique has played a seminal role in the study of lung mechanics since its establishment as a practical methodology in the mid-1980s [12]. Alveolar pressure relative to atmospheric pressure in an open-chest animal provides a direct measure of the pressure across the lung tissue, while alveolar pressure relative to the pressure at the airway opening equals the pressure drop along the airways. The alveolar capsule thus allows the pressure drop across the lung to be partitioned into the pressure losses due to air flow resistance along the airways and the remainder which is due to the elastic and dissipative forces acting across the lung tissue. In its usual manifestation, the alveolar capsule technique requires that the chest wall be opened and widely retracted. However, a capsule can also be installed through a hole dissected between two ribs [17], giving the pressure across both the lung and chest wall tissues together.

2.2.5   Flow transducers

The measurement of gas flow ranks alongside the measurement of pressure in its importance to the study of lung mechanics. Indeed, measuring flow (or, equivalently, volume) of air during a forced expiration is required for the calculation of FEV_1 and FVC, two of the most important diagnostic parameters in pulmonary medicine (Section 1.3).

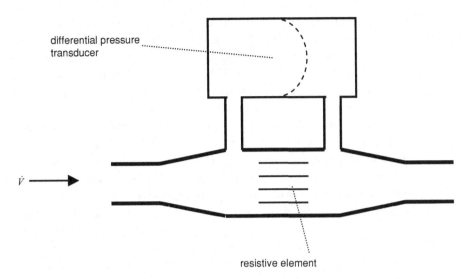

differential pressure transducer

$\dot{V}$

resistive element

Figure 2.11 The pneumotachograph.

The most commonly used device for measuring gas flow in respiratory applications is the *pneumotachograph*, which is a calibrated resistance (R) across which a differential pressure is measured (Fig. 2.11). Pneumotachographs are designed so that the pressure drop (ΔP) across the resistance is linearly proportional to the flow ($\dot{V}$) through it over the operating flow range. That is,

$$\Delta P = R\dot{V}. \qquad (2.8)$$

However, it is possible to utilize a pneumotachograph outside its linear range by using a computer to invert its pressure-flow relationship, should this become nonlinear.

The frequency response of a pneumotachograph depends on the construction of its resistive element. Some elements consist of a honeycomb arrangement of conduits, while others consist of a wire screen. The honeycomb type are less likely to become blocked by secretions, but have a poorer frequency response than the screen type. Either type should be heated to above body temperature during prolonged use to avoid breath condensate settling on the resistive element and changing its resistance (and hence altering the calibration of the device). Pneumotachographs are under-damped low-pass transducers. They can have a reasonably good frequency response out to about 20 Hz, after which the amplitude response increases to reach a peak at the resonant frequency of about 70 Hz. This resonance is due to the interaction between the compressibility and the mass of the gas inside the pneumotachograph and the tubing connecting it to its differential pressure transducer [1]. The connecting tubing should be as short as possible because the frequency response degrades rapidly with increasing length. If the frequency response is accurately characterized, however, it can be digitally corrected to a significant extent. This can give the pneumotachograph an effectively flat frequency response out to more than 100 Hz [18].

Another consideration for pneumotachographs concerns their *dynamic common-mode rejection* characteristics. If the two ports of a pneumotachograph are subjected to the same nonzero pressure, so that flow remains zero, the device should register a flow of zero. This will only be the case, however, if the common-mode rejection ratio of the differential transducer connected to the two ports is sufficiently high. Modern differential transducers generally perform well in this regard under static conditions (i.e. when the applied pressure is held constant). What tends to be more of a problem, however, is the dynamic common-mode rejection behavior that pertains when the applied pressure oscillates, again in the absence of flow. The problem arises because, in order for the two sides of the pressure transducer to register a change in common pressure, tiny volumes of gas must flow into and out of each lateral port to pressurize the gas reservoirs in the transducer either side of its pressure-sensitive element (Fig. 2.6B). If the time-constants associated with this process are different, which can happen when the physical dimensions of the transducers are not symmetrical with respect to the two ports, then a transient non-zero differential pressure will occur that will be recorded as a changing flow. Digital compensation methods can improve the situation to a certain extent [19], although for human-sized pneumotachographs dynamic common-mode rejection does not become problematic until frequencies far exceed those of normal breathing. However, dynamic common-mode rejection can be problematic in small-bore pneumotachographs that are sometimes used with very small animals [20] because the *input impedance* of the pressure transducer may no longer be large compared to that of the pneumotachograph itself. When this is the case, a significant fraction of the gas entering the pneumotachograph at any instant may be required to pressurize the transducer instead of flowing into the lungs of the animal.

Although the resistive pneumotachograph is the mainstay of flow measurement for respiratory applications, other devices have been used. Ultrasonic transducers based on differences in time-of-flight of sound propagating into and out of the direction of flow have an excellent frequency response and avoid the problems of a resistive element becoming clogged with secretions [21]. Devices based on the rate of cooling of a heated wire are also in use for respiratory applications [22].

2.2.6   Volume measurement

Changes in the volume of gas entering or leaving the lungs via the mouth are readily determined by a variety of means. Volume change measured in this way is almost equal to volume change within the lung, the difference being only a small amount of gas compression or expansion (usually less than 1%) due to the difference between alveolar and atmospheric pressures, plus small differences in the fluxes of oxygen and carbon dioxide across the blood-gas barrier. The traditional approach to measuring volume change at the mouth is to use a device known as a *spirometer*, which is a variable-volume gas reservoir from which a subject breathes. The walls surrounding the reservoir are designed to move as the reservoir changes volume in a way that can be tracked and recorded (the simplest realization of this device is to use a rigid container with its open end inverted over a bath of water, so the height of the container above the surface is proportional to the volume of gas it contains). Conventional spirometers are simple

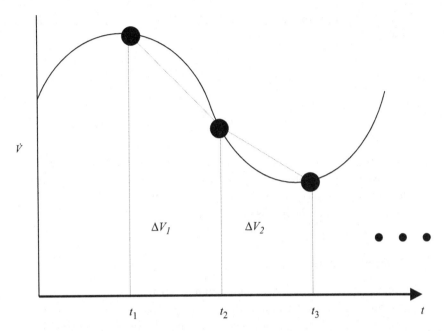

Figure 2.12 Trapezoidal integration of flow involves connecting adjacent flow points by straight lines and summing the areas (ΔV_i) under each trapezoid (adapted with permission from [25]).

and accurate devices, but they tend to be bulky and have limited frequency response. Alternatively, changes in lung volume can be estimated plethysmographically from changes in the volume of the body trunk. *Inductance plethysmography* [23] measures changes in the areas enclosed by a pair of elastic belts placed around the thorax and abdomen. This approach is convenient, but it assumes that changes in lung volume can be characterized in terms of only two mechanical degrees of freedom. A more accurate although more expensive and less convenient approach is *optical plethysmography*, which uses cameras to track the three-dimensional positions of numerous points on the body surface [24].

Volume can also be determined by *integrating* flow with respect to time. Before the advent of the modern laboratory digital computer, integration was typically achieved electronically by allowing a voltage proportional to flow to charge a capacitor (and before that it was not uncommon for the flow signal to be plotted on graph paper, cut out with scissors, and weighed!) Nowadays, integration is performed digitally on a computer. This can be achieved using a variety of algorithms that differ in the accuracy with which they approximate the true area under the flow curve. For most respiratory applications, the simple *trapezoidal rule* suffices and is performed as follows (Fig. 2.12). A digitized flow signal consists of a series of data points $\{\dot{V}_1, \dot{V}_2, \dot{V}_3, \ldots\}$ separated by equal time intervals Δt. Provided Δt is small enough, the flow curve between adjacent points can be well approximated by a straight line. The area, ΔV, under the trapezoid defined by the adjacent points $\dot{V}_1$ and $\dot{V}_2$ is

$$\Delta V \approx \frac{\dot{V}_1 + \dot{V}_2}{2} \Delta t. \tag{2.9}$$

The area V under the entire $\dot{V}$ curve is then simply the sum of all the individual ΔV_i as follows:

$$V \approx \sum_{i=1}^{n} \Delta V_i$$

$$= \left(\frac{\dot{V}_1}{2} + \dot{V}_2 + \dot{V}_3 + \cdots + \dot{V}_{n-1} + \frac{\dot{V}_n}{2} \right) \Delta t. \tag{2.10}$$

The key point is that Δt should be small enough so that the combined error accrued in approximating the true curve by a series of straight lines is negligible. This can be tested for in any particular situation by integrating the flow signal using progressively smaller values for Δt until the results do not change.

2.2.7   Plethysmography

Although tracking changes in lung volume is relatively straightforward, determining the absolute volume of gas in the lungs is another matter. It is impossible to measure this volume directly because the lungs cannot be emptied completely; even at the end of a maximal expiratory effort, more than a liter of air remains. Although it is possible, in principle, to directly measure quantities such as residual volume, functional residual capacity, and total lung capacity (Fig. 1.2) by imaging the thorax in three dimensions [26], this is not usually a practical solution. The volume of thoracic gas (V_{TG}) can, however, be inferred using *whole-body plethysmography*, which complements the measurement of forced expiratory flows as the other major methodology currently in clinical use for assessing lung function in patients [27].

A body plethysmograph is a rigid-walled container having a volume of about 500 L, inside which a subject sits. The plethysmograph functions to monitor changes in the volume of the subject's entire body, which can be achieved in either of two ways. One approach is to use a completely sealed box so that when the body volume increases by an amount ΔV, the volume of the air in the box around the body decreases by the same amount. This causes a corresponding change in box pressure (P_{box}). The analysis of this situation is based on *Boyle's law*, which states that for a given number of gas molecules isolated within a container at an initial volume V_1 and pressure P_1, when the gas is compressed (or expanded) to a new volume V_2 and pressure P_2 then

$$P_1 V_1 = P_2 V_2. \tag{2.11}$$

Putting this law in the context of the plethysmograph, knowing the initial volume of air in the box (V_{box}), one can determine ΔV as

$$1000 V_{box} = (1000 + P_{box})(V_{box} + \Delta V) \tag{2.12}$$

where the factor of 1000 corresponds to a pressure of one atmosphere in units of cmH_2O, and pressures are measured relative to atmospheric in the same units. Alternatively, the

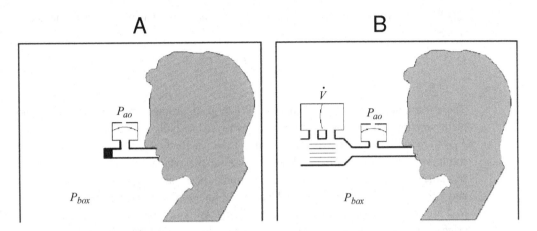

Figure 2.13 Body plethysmograph used to measure A) V_{TG}, and B) R_{aw}. P_{ao} is airway opening pressure, P_{box} is pressure inside the plethysmograph, and $\dot{V}$ is mouth flow (adapted with permission from [25]).

box can be allowed to vent either to a spirometer or to the atmosphere through a pneumotachograph so that the air displaced by a change in body volume is measured directly.

V_{TG} is determined by having a subject inside a body plethysmograph make breathing efforts against a closed airway while the changes in airway opening pressure (ΔP_{ao}) are measured (Fig. 2.13A). At the same time, the changes in the total volume of the subject (ΔV) are measured as described above. Provided that the only compressive part of the body is the gas in the lungs, ΔV is equal to the compression of the thoracic gas. Again applying Boyle's law gives

$$V_{TG} \approx \frac{-1000\Delta V}{\Delta P_{ao}}. \tag{2.13}$$

The other major use of body plethysmography is in the measurement of *airway resistance* (R_{aw}). Here, the subject breathes in a rapid shallow panting fashion while respiratory flow ($\dot{V}$) is measured at the mouth with a pneumotachograph simultaneously with P_{box} and P_{ao} (Fig. 2.13B). During this maneuver, a negative pressure (relative to atmospheric) occurs in the alveolar gas in the lungs during inspiration, and the converse during expiration. These changes in alveolar pressure (P_A) cause corresponding changes in V_{TG} according to Boyle's law as explained above. The changes in V_{TG} in turn cause equal and opposite changes in the volume (ΔV) of the plethysmographic gas, quantified from the resulting changes in P_{box} via Eq. 2.12. Having already determined V_{TG} as described above (Eq. 2.13), the only unknown quantity that remains is P_A (relative to atmosphere). This can be solved for, again via Boyle's law, as

$$P_A \approx \frac{-1000\Delta V}{V_{TG}}. \tag{2.14}$$

Finally, since the pressure drop along the pulmonary airways is now known, one determines

$$R_{aw} = \frac{P_{ao} - P_A}{\dot{V}}. \tag{2.15}$$

There is a complicating factor, however. When gas is inspired into the lungs from the plethysmograph it becomes humidified and heated to body temperature. This causes V_{TG} to increase by more than the volume of the gas that passed through the lips. The result is an extra increase in P_{box} during inspiration that cannot be separately determined from the change in P_{box} due to R_{aw}. Conditioning the air in the box to body temperature and humidity mitigates this effect. Also, plethysmographic measurements of V_{TG} and R_{aw} are based on the assumption that the lungs are a homogeneously ventilated (i.e. effectively single-compartment) system. This works well for normal lungs, but may fail rather significantly in severely diseased lungs [28].

2.3   Experimental scenarios

The various pressures, flows, and volumes described above each contain information pertaining to the mechanical functioning of the lungs. Indeed, the current basis of much of diagnostic pulmonary medicine is provided simply by the measurement of flow at the mouth during a forced expiration (Section 1.3). However, the real investigative potential of these signals is only realized when they are measured in combination. There are numerous possible such combinations, one of which is illustrated in Fig. 2.14. Here a supine intubated subject is connected to a mechanical ventilator through a pneumotachograph for the measurement of flow. A lateral tap between the pneumotachograph and the subject connects to a gauge pressure transducer for the measurement of pressure at the airway opening. An esophageal balloon provides esophageal pressure as a surrogate for pleural pressure. In this scenario, the pressure difference across the lungs (from the airway opening to the pleural surface) is reflected in the flow and volume of gas entering the lungs.

A related measurement scenario would be to have the subject instrumented for the measurement of the three signals shown in Fig. 2.14, but this time with the subject breathing spontaneously through a mouthpiece. Now inspiration is produced by negative pressure around the pleural surface rather than positive pressure at the airway opening.

In anesthetized laboratory animals it is possible to open the chest to expose the lungs, glue an alveolar capsule to the pleural surface, and use it to monitor alveolar pressure while the animal is receiving mechanical ventilation. Relative to the pressure at the airway opening, this gives the pressure drop across the pulmonary airways, which can then be related to airway flow. These are but a few examples of the numerous ways lung mechanics data can be collected. How these data are utilized to gain an understanding of lung mechanics itself is the subject of the rest of this book.

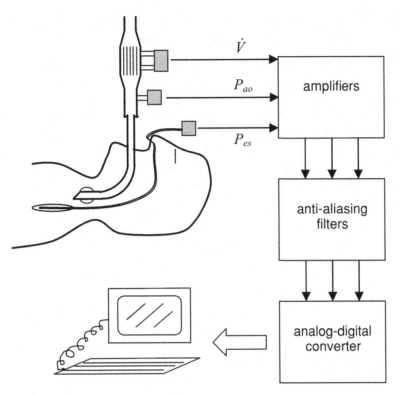

Figure 2.14 Measurement setup for assessment of pulmonary mechanics in a mechanically ventilated patient showing gas flow ($\dot{V}$) measured with a pneumotachograph, airway opening pressure (P_{ao}) measured through a lateral tap, and esophageal pressure (P_{es}) measured with a balloon catheter placed in the esophagus.

Problems

2.1 You are going to use a pneumotachograph to measure air flow at the mouth of exercising subjects who may generate flows of up to ± 10 L.s^{-1}. You need to measure these flows with an accuracy of 5 mL.s^{-1}. The differential pressure transducer to be used to measure the pressure drop across the pneumotachograph has a manufacturer-specified resolution of 1% of full scale. Assuming that full scale on the pressure transducer corresponds to the maximum flow magnitude to be measured, how many bits are required in an analog-digital converter to allow the transducer to be used to its full resolution?

2.2 If a sine wave of frequency 6 Hz is sampled at 10 Hz, what frequency do the sampled points appear to have?

2.3 Pressure is measured via a lateral tap in a conduit of circular cross-section through which air (density 1.2×10^{-3} g.mL^{-1}) flows at a constant rate. The diameter of the conduit is 1 cm and the resistance of the conduit downstream of the lateral tap is 10 cmH$_2$O.s.mL^{-1}. At what flow of gas (mL.s^{-1}) will the pressure at the lateral tap equal atmospheric pressure if the flow velocity profile is square? What flow will result in this condition if the velocity profile is parabolic?

2.4   Measurement of thoracic gas volume using plethysmography requires a subject to pant quickly against an occluded airway (Fig. 2.13A). This rests on the assumption that the pressure changes generated within the alveolar spaces of the lung are transmitted instantaneously up to the pressure transducer placed immediately behind the point of occlusion. This requires effectively instantaneous movement of small volumes of gas to and from the alveolar regions of the lung to the mouth. In severe obstructive lung disease, however, the airways may be so narrowed that alveolar and mouth pressures become disparate. What effect would this have on the value of thoracic gas volume obtained?

3   The linear single-compartment model

3.1   Establishing the model

3.1.1   Model structure

The first step in constructing a model of any physiological system is to decide on the structure of the model; how may independent components will it have, what do these components represent, and how are they connected to each other? There is no universally applicable answer to any of these questions, so one must be guided by whatever factors pertain to the situation at hand.

In addressing the question of how to model lung mechanics in the simplest possible terms, it is intuitively rather obvious that the lung can be viewed as being like a balloon sealed over the end of a pipe (Fig. 1.4). This model has a clear anatomical analogy to a real lung; the pipe represents the conducting airways and the balloon represents the elastic parenchymal tissue. There is also a functional analogy; the balloon can be inflated and deflated through the pipe in the same way that a lung inspires and expires. It goes without saying, of course, that a real lung is vastly more complicated than this simple construct. On the other hand, it is difficult to see how one could devise a simpler model that has any useful resemblance to the real thing. A tour of the universe of lung models, therefore, starts with the single-compartment model.

Although the balloon-on-a-pipe configuration shown in Fig. 1.4 is commonly how the lung is viewed, from now on we will be considering the representation shown in Fig. 3.1. The models in Figs. 1.4 and 3.1 are mathematically equivalent because the state of either is uniquely specified at any point in time by a single physical variable, namely the volume of gas in the elastic compartment. Accordingly, we say that these models have a single mechanical degree of freedom. The representation in Fig. 3.1, however, has the advantage of being readily expandable to incorporate more detailed descriptions of lung tissue, as will become apparent in later chapters. The alveolar compartment in this representation consists of a pair of telescoping canisters connected to each other by a spring which becomes stretched as the compartment volume (V) increases. The tension generated in the stretched spring produces an elastic pressure (P_{el}) inside the compartment. P_{el} is what makes the compartment return to its original volume when the inflating pressure is removed. This mimics the passive expiration that occurs in a lung when the respiratory muscles relax at the end of an inspiration. Some lungs are

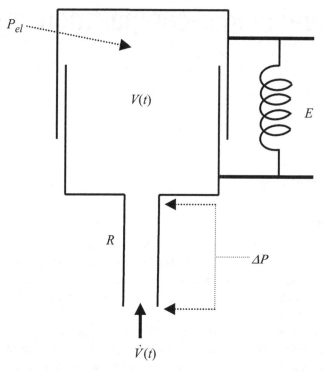

Figure 3.1 The single-compartment linear model of the lung. The compartment consists of a pair of telescoping canisters connected to each other by a spring with spring constant E. The single conduit serving the compartment has a flow resistance R. The elastic recoil pressure (P_{el}) inside the compartment is linearly related to the volume (V) of the compartment. The pressure difference (ΔP) between the two ends of the conduit is linearly related to the flow of gas ($\dot{V}$) through it. Under dynamic conditions, such as during breathing, V and $\dot{V}$ are functions of time (t).

more easily inflated than others, so by choosing the appropriate stiffness for the spring, the model can be made to represent a particular lung.

A single conduit connects the compartment of the model in Fig. 3.1 to the outside world. It takes a certain pressure (ΔP) to drive a given flow of gas through this conduit, in the same way that pressure is required to drive flow through the pulmonary airways. In other words, the conduit has a certain resistance to flow. This resistance is large (i.e. it takes a lot of pressure to produce a small flow) if the conduit is long and narrow, and small if the conduit is short and wide. By choosing the dimensions of the conduit appropriately, it can be given a resistance similar to that of any particular set of pulmonary airways.

3.1.2   The equation of motion

Deriving the equation of motion of the model in Fig. 3.1 requires certain assumptions about the elastic properties of the spring and the flow resistance of the conduit. These are called the *constitutive properties*.

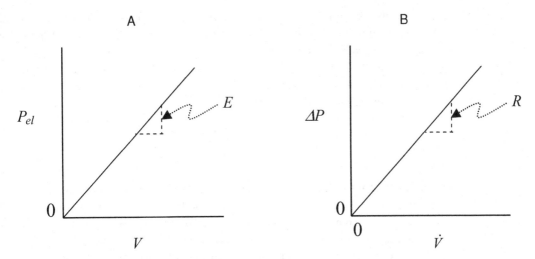

Figure 3.2 Simple constitutive properties for the linear single-compartment model. A) When elastance (E) is constant, the relationship between P_{el} and V is linear. B) When resistance (R) is constant, the relationship between ΔP and $\dot{V}$ is linear.

First, consider the elastic properties of the alveolar compartment that determine how P_{el} is related to V. If the spring is *Hookean* then its tension increases linearly as it is stretched past its relaxed length, so P_{el} inside the compartment increases linearly with V (Fig. 3.2A). We shall see in a later chapter just how well this represents the behavior of a real lung, but for now this assumption allows us to characterize the relationship between P_{el} and V in terms of a single number, the slope E. That is,

$$P_{el} = EV \tag{3.1}$$

where we have assumed that V is measured relative to the volume in the compartment when the spring is completely relaxed (equivalent to functional residual capacity, Fig. 1.2). E is termed *elastance* and is a measure of how difficult it is to inflate the alveolar compartment. Its inverse, *compliance*, is more commonly encountered in clinical applications and is a measure of how easily the compartment is inflated.

The flow resistance of the airway conduit can be characterized in a similarly simple fashion by assuming that the pressure difference (ΔP) between the proximal and distal ends of the conduit increases linearly with flow (Fig. 3.2B). That is,

$$\Delta P = R\dot{V} \tag{3.2}$$

where R is defined as the *resistance* of the conduit, and $\dot{V}$ is the derivative of V with respect to time. Again, it is important to bear in mind, as we develop these equations, that there is no such thing as a perfectly straight line in the real world. The relationships depicted in Fig. 3.2 are merely idealizations that enjoy the particular advantage of being characterized by single parameters, namely their slopes. Whenever the lung is characterized as having a particular value of R or a particular value of E, the linear relationships embodied in Fig. 3.2 are automatically invoked. The key issue is thus not whether these relationships are correct, but rather whether they suffice for the purpose at hand, and that depends on the purpose.

The total pressure across the model, from the entrance of the conduit to the outside of the compartment wall, is the sum of the pressure drop across the conduit and the pressure in the compartment (Fig. 3.1). That is,

$$P = P_{el} + \Delta P$$
$$= EV + R\dot{V}. \qquad (3.3)$$

It should be noted that Eq. 3.3 holds regardless of whether P is applied at the entrance to the airway or at the outside of the elastic compartment. In the former case, P is positive relative to atmospheric pressure, as would be the case for a patient receiving mechanical ventilation through an endotracheal tube. By contrast, a spontaneously breathing subject applies a negative P around the outside surface of the lungs by contracting the respiratory muscles. The P in Eq. 3.3 refers only to the difference in pressure between the airway opening and the pleural surface, irrespective of how this pressure is generated.

At this point, a comment about notation is in order because we have been denoting volume and flow by V and $\dot{V}$, respectively, as is the physiological convention. But now the worlds of physiology and applied mathematics are about to intersect, and we face a potential source of confusion. The convention in applied mathematics is to denote the time derivative of a quantity by placing a dot over its symbol. In this sense, $\dot{V}$ is the time derivative of V, and indeed in the case of the model depicted in Fig. 3.1 this is entirely true. For a real lung, however, it is not true. When ambient air is inspired, it becomes heated and humidified to match the conditions inside the lungs. This adds extra volume and results in thoracic expansion being slightly greater than integrated mouth flow. Gas compression within the lungs, necessary to produce the mouth-to-alveolus pressure differences that drive flow along the airways, also results in thoracic volume changes that are not immediately matched by flow at the mouth. Finally, the rate at which oxygen moves from the lungs into the pulmonary blood is generally slightly greater than the rate at which carbon dioxide is excreted from the blood into the lungs. This again contributes a small amount to changing gas volume within the lungs independent of gas flow entering the mouth. There is no question that the bulk of the change in lung volume comes from mouth flow, but it is important to realize that the two are not precisely equal in real life, even though we might make them equal in a model.

The equality of $\dot{V}$ and rate of change of lung volume for the model in Fig. 3.1 means that Eq. 3.3 is a first-order differential equation, the *equation of motion of the single-compartment linear model*. The *variables* of the model are P, V, and $\dot{V}$, and its *parameters* are E and R. It is a linear model because if one was to make a plot of the *dependent variable*, P, versus either of the *independent variables*, V or $\dot{V}$, when everything else is kept constant, the result would be a straight line.

Historically, Eq. 3.3 has had such importance in the field of lung mechanics that one sometimes sees it referred to as the "equation of motion of the lung," but it is important to remember that this is not correct. Equation 3.3 is simply the equation of a model, and

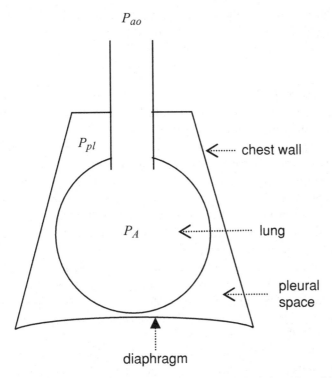

Figure 3.3 The lungs are normally encased in the thoracic cavity, which is bordered by the chest wall (composed of the ribs, sternum, spine, and associated respiratory muscles) and the diaphragm. The lungs are separated from these structures by the pleural space. Pleural pressure (P_{pl}) is the pressure within this space. Transpulmonary pressure (P_{tp}) is the difference between airway opening pressure (P_{ao}) and P_{pl}.

no model can ever behave exactly like the real thing, regardless of how useful it might be.

Equation 3.3 requires a slight modification before it can be applied to a real lung. In an intact subject, the lungs are encased in the thorax (Fig. 3.3) and thus are surrounded by the pressure in the pleural space (P_{el}), so P needs to be referenced to P_{pl}. The difference between the pressure at the airway opening (P_{ao}) and P_{pl} is known as *transpulmonary pressure* (P_{tp}). In most investigational situations in human subjects, esophageal pressure (P_{es}) is used as a surrogate for P_{pl} (Section 2.2.3). Consequently,

$$P_{tp} = P_{ao} - P_{es}. \tag{3.4}$$

It makes sense to reference V in Eq. 3.3 to functional residual capacity (Fig. 1.2) because this volume corresponds to the natural resting configuration of the lungs. In other words, V is defined as being zero at functional residual capacity even though the lungs still contain a significant volume of gas (Section 2.2.7). However, P_{tp} is not zero at functional residual capacity because P_{pl} is slightly negative in a subject whose respiratory muscles are relaxed and whose glottis is open (without this negative pressure, which is typically about 3 cmH$_2$O in magnitude, the lungs would collapse almost completely).

Equation 3.3 therefore requires the addition of an offset pressure (P_0) to account for the fact that P_{tp} is non-zero when V and $\dot{V}$ both are zero. Thus, when Eq. 3.3 is placed in the context of a lung inside an intact thorax, it becomes

$$P_{tp} = E_L V + R_L \dot{V} + P_0 \qquad (3.5)$$

where E_L and R_L specifically denote the elastance and resistance, respectively, of the lung. The single-compartment linear model of the lung in the intact human or animal is thus fully characterized by the three parameters E_L, R_L, and P_0.

P_0 is usually required in Eq. 3.5 even if the chest wall is open and widely retracted, as may be the case during invasive experimentation in anesthetized animals or even in humans during surgery [29]. This is because it is necessary to prevent the lungs from collapsing completely by applying a *positive end-expiratory pressure* (PEEP) of a few cmH$_2$O to the airway opening. Under these conditions, V is defined as being zero at the end of an expiration against the set level of PEEP, so P_0 represents the PEEP level itself. Furthermore, P_0 may be augmented when the duration of expiration is unusually short or when the lungs take an excessive amount of time to empty. Under these conditions, the lungs may not have enough time to empty to their elastic equilibrium volume, achieved when P_{el} balances PEEP, before the next inspiration starts. The extra volume left in the lungs at the end of expiration causes P_{el} to be elevated above resting P_{tp} by an amount known as *intrinsic* PEEP [30].

Measuring P_{ao} is easy to do; all that is required is a lateral pressure tap in a mouthpiece or a section of ventilator tubing (Section 2.2.2). Measuring P_{tp}, however, means either measuring P_{pl} directly, opening the chest so that P_{pl} equals atmospheric pressure, or else using some means such as an esophageal balloon (Section 2.2.3) to estimate P_{pl}. Neither situation is as convenient or noninvasive as simply measuring P_{ao} alone. However, when P_{tp} in Eq. 3.5 is replaced with P_{ao}, the parameters E and R pertain to the entire respiratory system (*rs*), which includes both the lungs and the chest wall. Accordingly, Eq. 3.5 becomes

$$P_{ao} = E_{rs} V + R_{rs} \dot{V} + P_0. \qquad (3.6)$$

The inclusion of P_0 is not strictly necessary here because P_{ao} is zero (relative to atmospheric pressure) when $\dot{V}$ and V are also both zero. Practically, however, it is not always easy to determine the precise point at which the lung has reached its elastic equilibrium volume, so the inclusion of P_0 in Eq. 3.6 is generally a good idea.

A third possible configuration for Eq. 3.5, again requiring the use of an esophageal balloon, is when P is replaced by P_{es}. Now the parameters E and R pertain to the chest wall (*cw*) alone, giving

$$P_{es} = E_{cw} V + R_{cw} \dot{V} + P_0, \qquad (3.7)$$

this time with P_0 representing resting P_{es}.

Yet another possibility offered by Fig. 3.3 is when P_{ao} is accompanied by a measurement of alveolar pressure (P_A), such as might be provided in animals by an alveolar capsule (Section 2.2.4). The difference between P_{ao} and P_A now gives a measure of the resistive pressure across the airways alone without any contribution from the lung

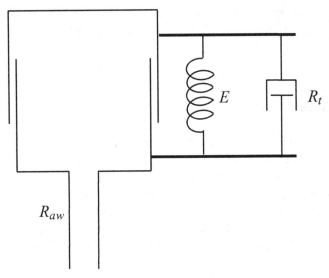

Figure 3.4 The single-compartment linear model of the lung showing the subdivision of resistance into its two components, R_{aw} and R_t.

tissues. This situation reduces Eq. 3.5 to

$$P_{ao} - P_A = R_{aw}\dot{V} \tag{3.8}$$

where R_{aw} is the resistance of the airways alone. P_A itself is a measure of the pressure across the lung tissue because there is no contribution from the chest wall when the alveolar capsule technique is applied in an open-chest animal. The elastic component of P_A under these conditions is therefore simply E_L.

Here, however, is where the model can be misleading. The depiction of the single-compartment model in Fig. 3.1 suggests that the only resistive component of the lung comes from the airways. In fact, the tissues of the lung also dissipate energy when they are stretched or distorted, and consequently they have their own resistance (R_t). The tissues thus obey an equation of motion similar to that of the entire lung. That is,

$$P_A - P_{pl} = E_L V + R_t \dot{V} + P_0. \tag{3.9}$$

The total resistance of the lung is thus the sum of the airway and tissue contributions:

$$R_L = R_{aw} + R_t. \tag{3.10}$$

Equation 3.5 is the sum of Eqs. 3.8 and 3.9. It turns out that in a normal lung during either regular breathing or conventional mechanical ventilation, R_{aw} and R_t both make significant contributions to R_L in Eq. 3.10 [14]. Accordingly, we need to modify the single-compartment model in Fig. 3.1 to include R_t, as shown in Fig. 3.4. Here is where the advantage of representing the model as a pair of telescoping canisters starts to become apparent. To represent R_t we merely need to add a resistive element known as a *dashpot* between the two horizontal moving components of the alveolar compartment.

3.2   Fitting the model to data

We have seen above how the linear single-compartment model can be used in different guises to relate measurements of P to corresponding measurements of $\dot{V}$ and V. The question now is, once the measurements are in hand, how do we use them to evaluate resistance and elastance in the relevant equation? In Eq. 3.5, for example, we want to choose the values of R_L, E_L, and P_0 so that the model behaves like a real lung. In other words, if the model were driven by the same $\dot{V}$ as measured entering a particular real lung, the model should predict a pressure similar to that measured in the lung. Of course, a predicted pressure can never be exactly the same as a measured pressure because that would require the model to behave exactly like the real lung, and that never happens in practice. We can, however, require that the predicted and measured pressures match each other as closely as possible.

3.2.1   Parameter estimation by least squares

Now we have to ask what "as closely as possible" means. Although there is no universally correct answer to this question, the approach that is invariably taken is to minimize the sum of the squared differences between the measured data and the corresponding model predictions. These differences are called *residuals*, so model fitting seeks to minimize the *sum of the squared residuals (SSR)*. Although *SSR* can never be zero because the model does not represent the real lung perfectly, there does exist a set of model parameter values for which *SSR* is a minimum. When this condition is met we say we have achieved the "best fit" of the model in the *least squares* sense, or that the model has been fit to the data by the method of least squares.

To illustrate how least squares fitting is achieved, suppose we want to fit the two-parameter model given by Eq. 3.3 to a set of measurements of P, V, and $\dot{V}$ (from now on we will use the generic terms P to denote pressure, R resistance, and E elastance, understanding that they might represent any of the corresponding particular quantities used in Eqs. 3.5–3.9). Clearly, if we choose some arbitrary value for R and multiply it by the measured $\dot{V}$ profile, then add the result to V multiplied by some arbitrary value of E, we will obtain a profile that probably bears little resemblance to the measured P. The trick is to choose values for R and E that cause the predicted P to match the measured P in the least squares sense, as defined above. This might sound like it could involve a lot of guessing until the right answer is hit upon, and for some model fitting problems this is essentially what has to be done. However, in the case of Eq. 3.3 we are in luck because R and E are linearly related to P, which means they appear as simple factors in their respective terms in the equation. Consequently, their best-fit values can be determined by a technique known as *multiple linear regression*. This gives an explicit formula for the best-fit values of R and E.

To see how multiple linear regression works, we begin by noting that its task is to minimize the expression

$$SSR = \sum_{i=1}^{N} [P_i - \hat{P}_i]^2$$

$$= \sum_{i=1}^{N} [P_i - EV_i - R\dot{V}_i]^2 \tag{3.11}$$

where P is the measured pressure, $\hat{P}$ is its prediction by Eq. 3.3, and the sum is taken over all N data points. If R is held fixed in Eq. 3.11 and only E is allowed to vary then SSR will be at its minimum when its derivative with respect to E is zero. That is,

$$\frac{\partial SSR}{\partial E} = -2 \sum_{i=1}^{N} V_i[P_i - EV_i - R\dot{V}_i]$$

$$= 0, \tag{3.12}$$

which rearranges to give

$$\sum_{i=1}^{N} V_i P_i = E \sum_{i=1}^{N} V_i^2 + R \sum_{i=1}^{N} V_i \dot{V}_i. \tag{3.13}$$

The same procedure can be applied with the roles of E and R reversed, giving

$$\sum_{i=1}^{N} \dot{V}_i P_i = E \sum_{i=1}^{N} V_i \dot{V}_i + R \sum_{i=1}^{N} \dot{V}_i^2 \tag{3.14}$$

Equations 3.13 and 3.14 constitute a set of two simultaneous equations, which can be solved to give unique answers for E and R as

$$R = \frac{\sum_{i=1}^{N} V_i^2 \sum_{i=1}^{N} \dot{V}_i P_i - \sum_{i=1}^{N} V_i \dot{V}_i \sum_{i=1}^{N} V_i P_i}{\sum_{i=1}^{N} V_i^2 \sum_{i=1}^{N} \dot{V}_i^2 - \left(\sum_{i=1}^{N} V_i \dot{V}_i\right)^2} \tag{3.15}$$

and

$$E = \frac{\sum_{i=1}^{N} V_i P_i - R \sum_{i=1}^{N} V_i \dot{V}_i}{\sum_{i=1}^{N} V_i^2}. \tag{3.16}$$

Equations 3.15 and 3.16 are exact solutions for the best-fit values of R and E.

This process can be extended to linear models that have an arbitrary number of parameters, but now it becomes convenient to cast the problem in terms of matrices. In the case of the three-parameter model given by Eq. 3.5, let n sequential measurements of the dependent variable P be written as a column vector thus:

$$\underline{P} = \begin{bmatrix} P_1 \\ P_2 \\ \vdots \\ P_n \end{bmatrix} \tag{3.17}$$

where a line under a symbol denotes it to be a vector. Similarly, we can represent the n corresponding measurements of the three independent variables as the matrix

$$\underline{\underline{X}} = \begin{bmatrix} V_1 \dot{V}_1 1 \\ V_2 \dot{V}_2 1 \\ \vdots \\ V_n \dot{V}_n 1 \end{bmatrix} \tag{3.18}$$

where a double underline denotes a matrix. The three parameters of the model are written as the vector

$$\underline{A} = \begin{bmatrix} E \\ R \\ P_0 \end{bmatrix}. \tag{3.19}$$

This formalism obviously extends to any number of independent variables and their corresponding parameters. The set of n equations expressing the model at each of the measurement points can then be written compactly as the matrix equation

$$\underline{P} = \underline{\underline{X}}\,\underline{A}. \tag{3.20}$$

Now the problem is to find the values of the elements in the vector $\underline{A}$. If we had the same number, m, of parameters as the number of measurements, n, then we could find $\underline{A}$ simply by pre-multiplying both sides of Eq. 3.20 by the inverse of the matrix $\underline{X}$. In the model fitting situation, however, m is less than n. Consequently, $\underline{\underline{X}}$ is not a square matrix and thus has no inverse. However, we can turn this situation into a problem of matrix inversion by pre-multiplying both sides of Eq. 3.20 by the transpose of $\underline{\underline{X}}$. This gives

$$\underline{\underline{X}}^T \underline{P} = \underline{\underline{X}}^T \underline{\underline{X}}\,\underline{A}. \tag{3.21}$$

When a matrix is multiplied by its transpose, the result is always a square matrix, which has an inverse provided its rows are independent. We can therefore premultiply Eq. 3.21 by the inverse of the square matrix $[\underline{\underline{X}}^T \underline{\underline{X}}]$ to obtain

$$\underline{A} = \left[\underline{\underline{X}}^T \underline{\underline{X}}\right]^{-1} \underline{\underline{X}}^T \underline{P}. \tag{3.22}$$

It turns out that this beautifully compact expression is also the least-squares solution for $\underline{A}$, making it the general formula for fitting a model to data by multiple linear regression.

3.2.2　Estimating confidence intervals

The fundamental assumption made in model fitting is that a model such as that in Eq. 3.20 takes into account all the *deterministic variation* in the dependent variable. This variation is assumed to be caused by well-defined processes governed by the laws of classical physics, and not influenced by chance. All uncertain aspects of the dependent variable are relegated to the remaining stochastic component, called *noise*. Equation 3.20 should thus really be written as

$$\underline{P} = \underline{\underline{X}}\,\underline{A} + \underline{N} \tag{3.23}$$

where $\underline{N}$ is a column vector of noise values and $\underline{A}$ is a vector of "true" parameter values. Consequently, when we use Eq. 3.22 we do not recover $\underline{A}$ exactly. Instead, we obtain an estimate of $\underline{A}$. Call this estimate $\underline{\hat{A}}$. Substituting Eq. 3.23 into Eq. 3.22 gives

$$\underline{\hat{A}} = \left[\underline{\underline{X}}^T \underline{\underline{X}}\right]^{-1} \underline{\underline{X}}^T (\underline{\underline{X}}\,\underline{A} + \underline{N})$$

$$= \left[\underline{\underline{X}}^T \underline{\underline{X}}\right]^{-1} \left[\underline{\underline{X}}^T \underline{\underline{X}}\right] \underline{A} + \left[\underline{\underline{X}}^T \underline{\underline{X}}\right]^{-1} \underline{\underline{X}}^T \underline{N}$$

$$= \underline{A} + \left[\underline{\underline{X}}^T \underline{\underline{X}}\right]^{-1} \underline{\underline{X}}^T \underline{N}. \tag{3.24}$$

Equation 3.24 shows that $\underline{\hat{A}}$ equals $\underline{A}$ plus a component containing $\underline{N}$. If the noise contained in $\underline{N}$ is uncorrelated and has a mean of zero then the expected value of its product with any deterministic quantity, such as the matrix $\underline{\underline{X}}$, is zero. This means that the expected value of the second term in Eq. 3.24 is zero, so the expected value of $\underline{\hat{A}}$ is A itself. $\underline{\hat{A}}$ is thus said to be an *unbiased* estimator of $\underline{A}$.

Now consider the matrix

$$[\underline{\hat{A}} - \underline{A}][\underline{\hat{A}} - \underline{A}]^T = \left[\Delta\underline{A}_1 \Delta\underline{A}_2 \cdots \Delta\underline{A}_m\right] \begin{bmatrix} \Delta\underline{A}_1 \\ \Delta\underline{A}_2 \\ \vdots \\ \Delta\underline{A}_m \end{bmatrix}$$

$$= \begin{bmatrix} (\Delta\underline{A}_1)^2 & \Delta\underline{A}_1\Delta\underline{A}_2 & \cdots & \Delta\underline{A}_1\Delta\underline{A}_m \\ \Delta\underline{A}_2\Delta\underline{A}_1 & (\Delta\underline{A}_2)^2 & \cdots & \Delta\underline{A}_2\Delta\underline{A}_m \\ \vdots & & & \\ \Delta\underline{A}_m\Delta\underline{A}_1 & \Delta\underline{A}_m\Delta\underline{A}_2 & \cdots & (\Delta\underline{A}_m)^2 \end{bmatrix}. \tag{3.25}$$

The elements on the leading diagonal (top left to bottom right) are, on average, proportional to the variances of the individual parameter estimates. The off-diagonal elements reflect the covariances between the various parameters. The entire matrix is known as the *covariance matrix* for $\underline{A}$. Writing $\underline{\hat{A}} - \underline{A}$ by rearranging Eq. 3.24 and substituting into Eq. 3.25 gives

$$[\underline{\hat{A}} - \underline{A}][\underline{\hat{A}} - \underline{A}]^T = \left[\underline{\underline{X}}^T \underline{\underline{X}}\right]^{-1} \underline{\underline{X}}^T \underline{N}\,\underline{N}^T \underline{\underline{X}} \left[\underline{\underline{X}}^T \underline{\underline{X}}\right]. \tag{3.26}$$

The only stochastic quantity in Eq. 3.26 is $\underline{N}$. If we make the assumption that the noise comprising $\underline{N}$ is uncorrelated, normally distributed, and has the same variance σ^2 at each data point, then on average the matrix $[\underline{N}\underline{N}^T]$ (the covariance matrix of the noise) is simply the identity matrix ($\underline{I}$) times σ^2. Equation 3.26 thus becomes

$$\left[\underline{X}^T\underline{X}\right]^{-1}\underline{X}^T\left(\sigma^2\underline{I}\right)\underline{X}\left[\underline{X}^T\underline{X}\right] = \left[\underline{X}^T\underline{X}\right]^{-1}\underline{X}^T\underline{X}\left[\underline{X}^T\underline{X}\right]\sigma^2$$
$$= \left[\underline{X}^T\underline{X}\right]\sigma^2. \tag{3.27}$$

The elements of the estimated parameter vector $\underline{\hat{A}}$ are thus normally distributed random variables with means equal to the corresponding true parameter values and standard deviations given by the square roots of the leading diagonal terms of the matrix $[\underline{X}^T\underline{X}]\sigma^2$. Of course, we do not know *a priori* what σ^2 is, but if the model has done a good job of describing the deterministic variation in the data, so that basically all that is left over is the noise, then σ^2 should be similar to the *mean squared residual (MSR)*. An unbiased estimate of σ^2 in this case is

$$\hat{\sigma}^2 = MSR$$
$$= \frac{SSR}{n-m}. \tag{3.28}$$

We can therefore take the square roots of the leading diagonal terms of the matrix $\hat{\sigma}^2[\underline{X}^T\underline{X}]$ as estimates of the standard deviations of the estimated parameter values. In order to use these standard deviations to obtain confidence intervals about the estimated parameter values, we need to multiply them by the appropriate value of the Student's t-distribution. This takes into account the uncertainty inherent in using $\hat{\sigma}^2$ as an estimate of σ^2. However, when the single-compartment model is fit to respiratory data, one is typically dealing with several hundred data points, so $n-m$ is a large number and the Student's t-distribution is close to the normal distribution.

An important final point that must be made about the multiple linear regression theory outlined above is that it is based on the assumption that the only noise in the system appears additively in the dependent variable, as expressed in Eq. 3.23. In reality, one also expects noise to be present in the independent variables V and $\dot{V}$ because, after all, they are experimental measurements too. If we denote this noise by the matrix $\underline{Q}$, then Eq. 3.24 becomes

$$\underline{\hat{A}} = \left[(\underline{X}+\underline{Q})^T(\underline{X}+\underline{Q})\right]^{-1}(\underline{X}+\underline{Q})^T(\underline{X}\underline{A}+\underline{N})$$
$$= \left[\underline{X}^T\underline{X}+\underline{X}^T\underline{Q}+\underline{Q}^T\underline{X}+\underline{Q}^T\underline{Q}\right]^{-1}\left[\underline{X}^T\underline{X}\underline{A}+\underline{X}^T\underline{N}+\underline{Q}^T\underline{X}\underline{A}+\underline{Q}^T\underline{N}\right]. \tag{3.29}$$

If the elements of $\underline{Q}$ are random uncorrelated noise, then the expected value of the product of $\underline{Q}$ with either $\underline{X}$ or $\underline{N}$ is zero. However, the expected value of the product of $\underline{Q}$ with itself is not zero, but rather equal to the covariance matrix of the noise. For simplicity we can assume this noise to have the same variance, q^2, at all time points, but this still leads to

$$\underline{\hat{A}} = \left[\underline{X}^T\underline{X}+q^2\underline{I}\right]^{-1}\underline{X}^T\underline{X}\underline{A} \tag{3.30}$$

which is not equal to $\underline{A}$. Thus, when there is noise in the independent variables, the parameter estimates produced by multiple linear regression are *biased*. This is very often the case in practice. It is therefore important to understand that the elegant theory outlined above frequently represents only an approximation to reality.

3.2.3   An example of model fitting

Figure 3.5 shows an example of how well the single-compartment model can describe respiratory mechanical data collected under the conditions illustrated in Fig. 2.14. The top three panels show flow, volume, and transpulmonary pressure, respectively, measured over a single breath in a mechanically ventilated patient. Also shown, in the third panel, is the fit to P_{tp} predicted by Eq. 3.5. Even though the fit is not perfect, it is nevertheless extremely good, accounting for the vast majority of the variation in P_{tp}. A measure of the goodness-of-fit of the model is provided by the *coefficient of determination (CD)*, which expresses the fraction of the variation in the data accounted for by the model according to

$$CD = 1 - \frac{SSR}{\sum\limits_{1}^{n}(P_i - \bar{P})^2} \tag{3.31}$$

where $\bar{P}$ is the mean value of the measured pressure signal. The denominator in Eq. 3.31 expresses the total variation in the dependent variable. CD can take a value anywhere from 1, which corresponds to a perfect fit, to 0, which means the model has no relation to the data whatsoever. The value of CD for the fit shown in Fig. 3.5 is 0.987, so the model in this case accounts for more than 98% of the variation in the data.

The value of R returned by the model fit shown in Fig. 3.5 is 7.1 cmH$_2$O.s.L^{-1} with a standard deviation estimated according to the theory outlined in Section 3.2.2 of 0.1 cmH$_2$O.s.L^{-1}. This gives a 95% confidence interval about the fitted value (approximately ± 2 standard deviations) of about 0.4 cmH$_2$O.s.L^{-1}, which is a very small range compared to the parameter value itself. The estimated value of E is similarly tight, with a best-fit value of 5.8 cmH$_2$O.L^{-1} and a 95% confidence interval centered on this value of about 0.4 cmH$_2$O.L^{-1}. Recall, however, that the estimation of these confidence intervals is predicated on the assumption that the residuals are random and uncorrelated. Inspection of the residuals in Fig. 3.5 shows this to be far from the case. In fact the residuals from this model fit (Fig. 3.5, bottom panel) are highly systematic, swinging either side of zero with a frequency somewhat greater than 1 Hz. These swings are caused by the heart bumping into the lungs each time it beats, and are known as *cardiogenic oscillations*.

One way to reduce the residuals is to ensemble-average a number of breaths, provided respiration is sufficiently regular and the cardiogenic oscillations have a random phase relationship with the breathing cycle. This is illustrated in Fig. 3.6A, which shows 60 s of data from the same subject as in Fig. 3.5. The individual breaths in these records were ensemble-averaged (overlaid and averaged point by point) to produce mean signals for V, $\dot{V}$, and P_{tp}, with much-reduced noise levels. Now Eq. 3.5 does an even better job of describing the data, accounting for 99.5% of the variance in the ensemble-averaged

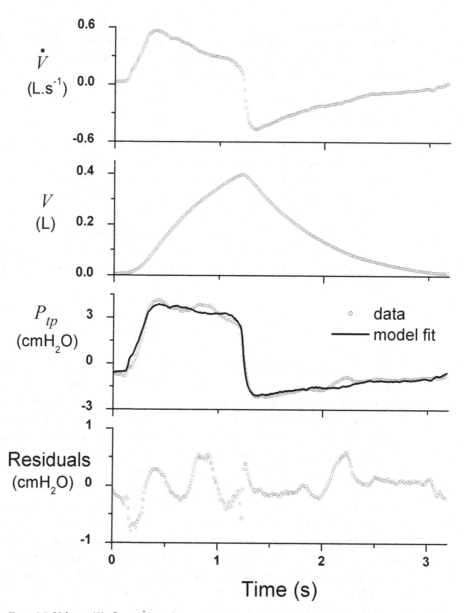

Figure 3.5 Volume (V), flow ($\dot{V}$), and transpulmonary pressure (P_{tp}) measured in a mechanically ventilated patient together with the fit provided by Eq. 3.5. The residuals are dominated by cardiogenic oscillations.

P_{tp} signal. Indeed, it is difficult to visually discern the fit from the data in Fig. 3.6B. Nevertheless, the residuals are still clearly systematic. In fact, they exhibit more extended runs either side of zero (Fig. 3.6B) than was the case before ensemble-averaging (Fig. 3.5).

Thus, although ensemble-averaging reduced the noise due to cardiogenic oscillations and other random sources of variation, and substantially reduced the magnitude of the

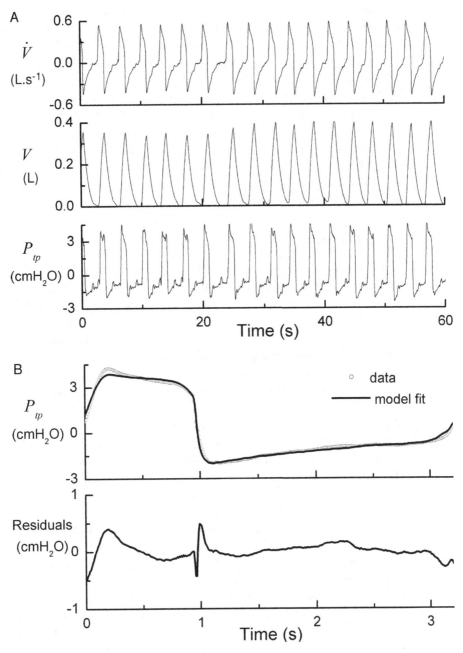

Figure 3.6 (A) Sixty s recordings of data from a mechanically ventilated patient. V was obtained by numerical integration of $\dot{V}$. P_{tp} is the difference between measurements of P_{ao} and P_{es}. (B) Fit of Eq. 3.5 to single-breath ensemble average of the 60 s data records, together with residuals.

residuals, it brought them no closer to having the desired characteristic of uncorrelated randomness. This is not an uncommon occurrence, so strictly speaking the elegant theory outlined in Section 3.2.2 is frequently inapplicable. However, this does not mean that all is lost. Certainly, it is not possible to assign confidence ranges to fitted parameter values in terms of statistical probabilities unless the residuals have the required characteristic of randomness. Nevertheless, one can still use the parameter standard deviations provided by regression theory as rationalized measures of how well the parameters are determined by the data. That is, a small standard deviation means that the parameter in question cannot be varied very much either side of its best-fit value without substantially degrading the quality of the model fit. Conversely, a large standard deviation means the parameter does not have much influence on how well the model fits the data. Such parameter sensitivity information is extremely valuable.

3.2.4 A historical note

A reader of the preceding sections of this chapter might be forgiven for assuming that fitting the single-compartment model to respiratory data is something that must be done by multiple linear regression and so had to wait for the widespread availability of digital computers. In fact, values for R and E were first calculated long before computers were invented. The theory of fitting models to data by least squares was well known at that time, but using it meant going through the above calculations by hand, which was simply not practical. Consequently, alternative methods were developed to estimate R and E. One of the first was the flow *interrupter method* in which a shutter is rapidly closed in front of the mouth while the pressure just behind the shutter is measured [31]. The sudden cessation of flow causes an equally sudden change in pressure (ΔP) measured behind the shutter that is taken to reflect the immediately preceding pressure drop along the pulmonary airways. Dividing ΔP by the value of $\dot{V}$ recorded just prior to interruption (Fig. 3.7A) then gives an estimate of R_{aw}.

An alternative manual approach to estimating E and R is the so-called *isovolume method* illustrated in Fig. 3.7B. Here, one locates two points within the same breath at the same volume, one during inspiration and one during expiration. According to the single-compartment model, P_{el} is the same at these two points, so the difference between the pressures at these points ($P_1 - P_2$) is equal to the sum of their respective values of $\dot{V}$ multiplied by R. In general, the value of R one obtains with the isovolume method depends on the lung volume at which the two points are taken [32].

A third manual method that became very widely used was that due to Mead and Whittenberger [33], who displayed P versus $\dot{V}$ on an oscilloscope to make a loop (Fig. 3.7C, top panel). They then subtracted from P a signal proportional to V, with the constant of proportionality manually adjusted to make the resulting plot as close to a single-valued curve as possible (Fig. 3.7C, bottom panel). The slope of this line then gives an estimate of R.

These pre-computer era approaches to parameter estimation have now been superseded by multiple linear regression, which has the advantages of convenience and robustness because all the data in the signals are utilized in the calculations instead of only a few

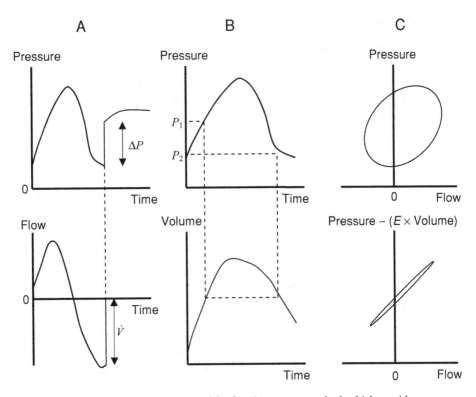

Figure 3.7 (A) Schematic representation of the flow interrupter method, which provides an estimate of R as the ratio of ΔP to $\dot{V}$. (B) The isovolume method, which uses points at equal volume to eliminate the contribution from E, leaving a pressure component determined only by R. (C) The Mead-Whittenberger method, which closes the P-$\dot{V}$ loop by subtracting from P a term proportional to V, leaving a relationship with slope equal to R.

points within each breath. However, the manual methods played very important roles in early investigations of respiratory mechanics [34] and were responsible for many of the insights upon which our current understanding of the subject rests.

3.3   Tracking model parameters that change with time

When Eq. 3.5 is fit to measurements of P, V, and $\dot{V}$ using multiple linear regression, it is assumed that R and E do not change over the period of data collection. Frequently, however, the lung is not at mechanical steady state while measurements are being made, in which case R and E are not constant. One way to deal with this situation is to fit the model only to data within a small time window, and repeat the fitting procedure as the window is moved along the data record. Of course, R and E must still be essentially constant over the duration of the window, but they can change from one window to the next. Obviously, the shorter the window, the better one is able to resolve rapid changes in R and E. On the other hand, shorter windows mean that fewer measurements of P, V, and $\dot{V}$ are used for each fit, so the estimates of R and E are more sensitive to noise.

3.3.1   Recursive multiple linear regression

A computationally efficient way of implementing a moving-window analysis is achieved by a *recursive* version of multiple linear regression. The essence of this algorithm is demonstrated by a simple example. Suppose you need to keep tabs on the mean value of a data stream. One way to do this is to wait for the first n values to come in and then calculate their mean as

$$\bar{x}_n = \frac{1}{n} \sum_{i=1}^{n} x_i. \tag{3.32}$$

This calculation can be repeated with the arrival of each new data point. Thus, after the $(n + 1)$th point we obtain

$$\bar{x}_{n+1} = \frac{1}{n+1} \sum_{i=1}^{n+1} x_i \tag{3.33}$$

and so on. However, this is a very inefficient way to proceed because all the calculations performed in Eq. 3.32 are repeated in Eq. 3.33. To save on this double work, one can re-express Eq. 3.33 as

$$\bar{x}_{n+1} = \frac{n}{n+1} \left(\frac{1}{n} \sum_{i=1}^{n} x_i \right) + \frac{1}{n+1} x_{n+1}$$

$$= \left(\frac{n}{n+1} \right) \bar{x}_n + \left(\frac{1}{n+1} \right) x_{n+1}. \tag{3.34}$$

In other words, the mean value of x after $n + 1$ points have arrived is simply the previous mean multiplied by a correction factor plus a scaled version of the new value of x. Consequently, none of the calculations performed in the determination of the previous mean have to be repeated. Equation 3.34 is the recursive formula for the arithmetic mean of a time series.

Recursive formulae have an additional very attractive feature in that they can be given a *fading memory*. The formula given in Eq. 3.34 has an infinite memory because it assigns equal importance to each value of x no matter how long ago it was measured. However, if $\bar{x}$ changes with time then its current value should be influenced only by the most recent values of x. A convenient way to achieve this is to weight the individual values of x in a manner that falls off exponentially with time into the past. The exponentially weighted mean of x is given by

$$\bar{x}_n = (1 - \lambda) \sum_{i=1}^{n} x_i \lambda^{n-i} \tag{3.35}$$

where λ is a constant with a value somewhere between 0 and 1. The time-constant of the fading memory in Eq. 3.35 is

$$\tau = \frac{-\delta t}{\ln(\lambda)} \tag{3.36}$$

where δt is the data sampling interval. The exponentially weighted mean also has a convenient recursive formulation as

$$\bar{x}_{n+1} = (1 - \lambda) \left(\lambda \sum_{i=1}^{n} x_i \lambda^{n-i} + x_{n+1} \right)$$
$$= \lambda \bar{x}_n + (1 - \lambda) x_{n+1}. \tag{3.37}$$

Adjusting the value of λ, the so-called *forgetting factor*, conveniently adjusts the length of the memory in the algorithm via Eq. 3.36. If λ is close to 1 then the algorithm has a long memory suitable for tracking values of $\bar{x}$ that do not change rapidly with time, while making λ closer to zero reduces the memory length and improves its ability to follow a rapidly changing $\bar{x}$, although at the expense of noise amplification.

Multiple linear regression can be formulated as a recursive algorithm with exponential memory in much the same way as for the exponentially weighted mean described above [5, 35]. Using recursive multiple linear regression to fit the single-compartment linear model of the lung to measurements of P, V, and $\dot{V}$ allows changing values of R and E to be tracked in time. The recursive form of multiple linear regression is

$$\hat{\underline{A}}_{k+1} = \hat{\underline{A}}_k + \frac{\underline{M}_{k+1} \underline{X}_{k+1} \left(y_{k+1} - \underline{X}_{k+1}^T \hat{\underline{A}}_k \right)}{\lambda + \underline{X}_{k+1}^T \underline{M}_k \underline{X}_{k+1}}$$

where

$$\underline{M}_{k+1} = \frac{1}{\lambda} \left[\underline{M}_k + \frac{\underline{M}_k \underline{X}_{k+1} \underline{X}_{k+1}^T \underline{M}_k}{\lambda + \underline{X}_{k+1}^T \underline{M}_k \underline{X}_{k+1}} \right]. \tag{3.38}$$

Here, $\hat{\underline{A}}_k$ is the vector of parameter values determined from the first k sets of measurements, $\hat{\underline{A}}_{k+1}$ is its update determined when the $(k+1)$th set of measurements arrives, $\underline{X}_{k+1}$ is the vector composed of the $(k+1)$th set of measurements, and $\underline{M}$ is the *information matrix* (corresponding to $[\underline{X}^T \underline{X}]^{-1}$ in Eq. 3.22) that also has to be updated at each step along with $\hat{\underline{A}}$ itself.

The recursive multiple linear regression formula (Eqs. 3.38 and 3.39) is a bit more complicated than the formula for the recursive mean (Eq. 3.37), but both share the same basic structure; the parameter estimates at each time step are given by scaled versions of the estimates at the previous time step plus correction terms proportional to the difference between the model prediction at the current step and the actual measurement. With multiple linear regression, however, there is the additional question of how to choose the initial parameter values. In the case of the recursive mean, the answer is easy; the estimate of the mean when $n = 1$ is simply the value of the first data point. But when two or more parameters have to be estimated simultaneously, there is not enough information in the first set of measurements to uniquely estimate parameter values. One way around this problem is to use conventional multiple linear regression to estimate parameter values using the first few sets of data, and then proceed recursively from then on. An alternative and more practical method is simply to set all the initial elements of $\hat{\underline{A}}$ to zero and all the leading diagonal elements of $\underline{M}$ to a very large number. Because the diagonal elements of $\underline{M}$ are proportional to the variances of the corresponding elements

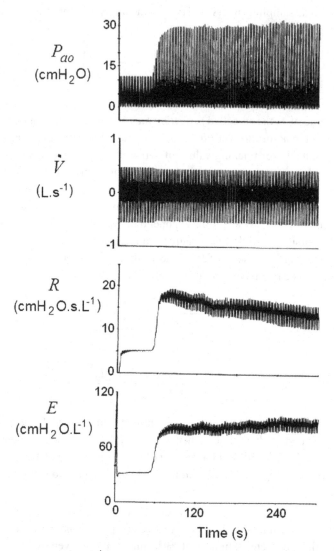

Figure 3.8 P_{ao} and $\dot{V}$ were measured in a dog injected at 60 s with a bolus of methacholine. Recursive multiple linear regression with a short memory time-constant was used to fit Eq. 3.5 to these data, producing the profiles of R and E shown (reproduced with permission from [35]).

of $\underline{\hat{A}}$, this has the effect of assigning negligible influence to the initial parameter values. Thus, the influence of the initial parameter values is very quickly forgotten as the subsequent data are processed through the algorithm [35, 36].

Figure 3.8 shows the time courses of R and E obtained when the single-compartment linear model (Eq. 3.5) was fit recursively to data obtained from a tracheostomized dog over a five-minute period when a bolus injection of methacholine (a smooth muscle agonist) was administered after one minute [35]. The animal was ventilated with a fixed tidal volume so $\dot{V}$ was virtually unaffected by the drug. However, the peak in

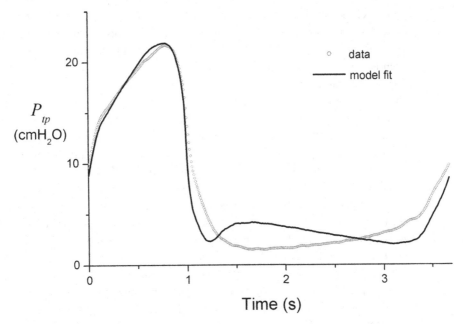

Figure 3.9 Ensemble-averaged P_{tp} signal from a mechanically ventilated patient together with the fit provided by Eq. 3.5.

airway pressure (P_{ao}) with each breath increased dramatically as the airway smooth muscle in the lungs reacted to the bronchial agonist. The profiles of R and E responded similarly. Of course, inspection of P_{ao} is enough to show that something drastic suddenly took place at the one-minute time point. However, during the subsequent course there was a gradual recovery in R offset by a slight progressive increase in E. These subtle differential effects on the resistive and elastic properties of the lungs are not readily apparent from visual inspection of P_{ao} alone; they are only revealed by tracking R and E individually.

3.3.2 Dealing with systematic residuals

Recursive multiple linear regression also allows us to take a somewhat novel approach to the calculation of confidence intervals about best-fit values of R and E. In the conventional approach described in Section 3.2.2, the noise in the data is assumed to arise only in the dependent regression variable (P) and to be random, uncorrelated, and Gaussian. Under these conditions, the parameter values provided by multiple linear regression are viewed as random variables themselves, with mean values equal to the true parameter values and variances provided by the covariance matrix (Eq. 3.27). In other words, confidence in the correctness of an estimated parameter value is expressed in terms of probability. Obviously, this is only appropriate if the above assumptions are satisfied. The problem is that they usually are not when Eq. 3.5 is fit to respiratory data. Figure 3.9 shows an example of such a situation. Here, the residuals are distributed about zero so systematically that it is difficult to see how the imperfections in the fit have any stochastic element to them at all. This presents a significant problem. If the probabilistic theory of confidence intervals developed in Section 3.2.2 is inapplicable, then how do we assign regions of confidence about the estimated values of R and E? Best-fit parameter values are of no use unless one has at least some idea of how far off the mark they might be.

The conventional way of dealing with systematic residuals is to try different (and usually more complicated) models until one is found that produces residuals with the desired character of random noise. One approach to improving the fit in Fig. 3.9 would therefore be to add higher-order polynomial terms in V and $\dot{V}$ to Eq. 3.5 [37]. The problem with this, however, is that it rapidly leads to an overwhelming number of uninterpretable parameters. We thus seem to be stuck between a rock and a hard place; the model needs to be simple to be manageable, yet we cannot invoke the powerful tools of statistical theory unless the model is allowed to become too complicated to interpret.

Of these two competing forces, retaining the ability to interpret a model in physiological terms is paramount. Models of lung mechanics are constructed with the aim of gaining insight about the lung's inner workings, and this is not achieved by constructing purely empirical descriptions of a particular data set. We thus come back to the two parameters R and E as descriptors of the overall dissipative and elastic properties of the lung, but face the question of how to gauge the uncertainty in their estimated values. One option is to stick with the classically estimated confidence intervals, but realize that since their interpretation as probabilities of correctness is now untenable, they should be viewed simply as indices of how sensitive the parameters are to variations in the data.

An alternative approach is to view systematic residuals as evidence that the parameters of the model vary across the data set. This can be appreciated by fitting the model to the data in short segments. The model will be able to mimic all the deterministic variation in the individual segments if they are short enough, but the estimated parameters will be different for each segment. Recursive multiple linear regression provides a convenient way to implement this idea.

Figure 3.10 shows the data used in Fig. 3.9 together with the fit achieved with the linear single-compartment model (Eq. 3.5) using recursive multiple linear regression with a memory time-constant of 0.1 s. Now the fit is indistinguishable from the data, but the corresponding R and E signals vary substantially. In fact, over significant portions of the data these parameters are negative, which is physically meaningless. However, these negative values correspond to regions of the data, principally at the end of expiration, that themselves change very little with time. Such regions contain little useful information about the parameters, as reflected in very large values of their respective diagonal terms in the covariance matrix $\underline{M}$ (Eq. 3.39). This provides objective justification for giving little credence to the corresponding estimates of R and E.

A convenient way to represent the recursive estimates of R and E in Fig. 3.10 is to construct histograms of their values across the time span of the data with the contribution of each value to its bin in the histogram weighted inversely by the magnitude of the corresponding diagonal element of the $\underline{M}$ matrix [5]. This provides a picture of

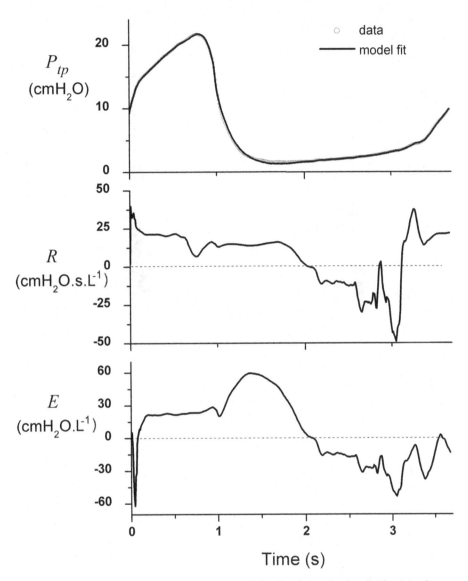

Figure 3.10 Time profiles for R and E obtained by fitting Eq. 3.5 to the data in Fig. 3.9 using recursive multiple linear regression with a memory time-constant of 0.1 s. Also shown is the fit to P_{tp} produced by the model.

the variation in each parameter that is tempered by the relative confidence one has in the estimates obtained at each time point.

Figure 3.11 shows examples of such information-weighted histograms calculated from upper airway pressure and flow data collected during sleep in normal subjects and subjects with obstructive sleep apnea [38]. The histograms can be characterized in terms of measures of central tendency (mean) and dispersion (standard deviation), and thus provide ranges for R and E. This represents an alternative to the classical method of

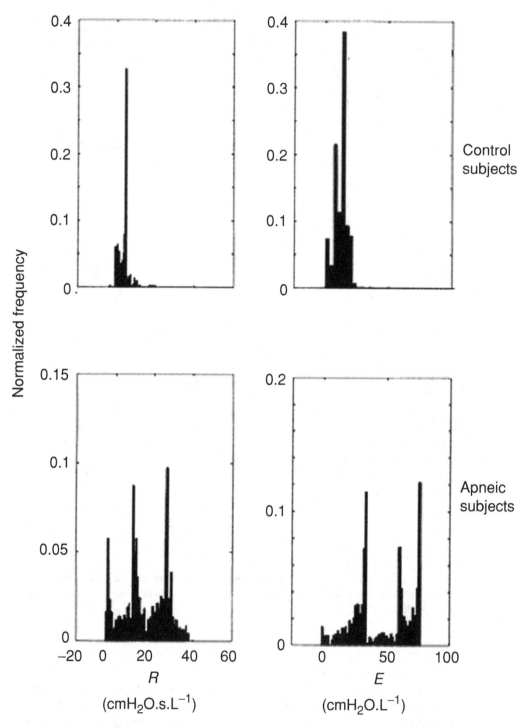

Figure 3.11 Information-weighted histograms for R and E from a normal sleeping subject (top) and a subject experiencing episodes of obstructive sleep apnea (bottom). Increased variations in R and E throughout the respiratory cycle are clearly evident in the apneic patient (reproduced with permission from [38]).

determining parameter confidence intervals and avoids statistical assumptions that are, in this case, not even close to being satisfied.

Problems

3.1 The derivation of the equation of motion of the single-compartment model of the lung (Eq. 3.3) assumes that the rate of change of lung volume is equal at all times to the flow of gas entering the mouth. Explain why this is not completely true for reasons related to (a) gas exchange, (b) gas compression, and (c) gas humidification.

3.2 Extend the diagram shown in Fig. 3.4 to include the chest wall.

3.3 From the perspective of the single-compartment model of the lung (Fig. 3.1), there is no functional difference between inflating the model using positive pressure at the airway opening and using negative pressure around the outside of the alveolar compartment. Satisfy yourself that this is so, and then think of extensions to the model that might cause this symmetry to be broken. Consider models consisting of two alveolar compartments connected either in series or parallel. Map these concepts onto a real lung in a living person. Does the symmetry still apply?

3.4 Show that Eq. 3.22 provides the least-squares estimate of the parameter vector of a linear model. (Hint: note that $SSR = [\underline{Y} - \underline{\underline{X}}\ \underline{A}]^{\mathrm{T}}[\underline{Y} - \underline{\underline{X}}\ \underline{A}] = \underline{Y}^{\mathrm{T}}\underline{Y} - \underline{Y}^{\mathrm{T}}\underline{\underline{X}}\ \underline{A} - \underline{A}^{\mathrm{T}}\underline{\underline{X}}^{\mathrm{T}}\underline{Y} + \underline{A}^{\mathrm{T}}\underline{\underline{X}}^{\mathrm{T}}\underline{\underline{X}}\ \underline{A}$. Differentiate SSR with respect to A and set the result equal to zero.)

3.5 Examine Eq. 3.24 and show that when the noise ($\underline{N}$) in a set of measurements ($\underline{Y}$) is correlated from one measurement to the next, Eq. 3.22 provides biased estimates of the model parameters.

3.6 The flow interrupter method for measuring airway resistance illustrated in Fig. 3.7A is based on the assumption that flow at the mouth is interrupted instantaneously. How will the estimate of resistance be affected if the valve used to interrupt flow takes a finite amount of time to close? (Hint: consider how the elastic recoil pressure within the lungs changes as the valve is in the process of closing.)

3.7 Derive Eq. 3.37.

3.8 The plots of recursively estimated R and E shown in Fig. 3.8 show breath-by-breath oscillations after the reaction to methacholine has taken place. These oscillations show that R and E vary throughout the breath. What might such variations be due to?

3.9 Figure 3.11 shows that information-weighted histograms for R and E are much more widely dispersed in sleeping subjects with obstructive sleep apnea (in whom the upper airways collapse and may become completely blocked temporarily) than in normal sleeping subjects. Why might this happen? (Hint: subjects with sleep apnea tend to snore loudly, particularly during inspiration.)

4   Resistance and elastance

Many common diseases of the lung involve alterations in lung mechanics. Being able to characterize these alterations is thus of great importance. Most research involving the model-based estimation of lung mechanics involves nothing more than what has just been described in the previous chapter. That is, the lung is assumed to behave like a single linear compartment characterized by only two parameters, resistance (R_L) and elastance (E_L). These two parameters are typically evaluated first under baseline or control conditions and then following whatever intervention is being investigated. This raises the question of how to interpret R and E, and their changes, in physiological terms. In this chapter we consider what these two parameters can tell us about events happening within the lungs.

We saw in the previous chapter (Fig. 3.4) that R_L is made up of two distinct components – one from the airways (R_{aw}) and one from the lung tissues (R_t). This raises questions about how these two components arise. How, for example, could the airway tree, with all its structural complexity, give rise to a particular value of R_{aw}? Why is it that when energy is imparted to the tissues during lung inflation, some portion of this energy is dissipated to give rise to R_t while the remainder is stored to be reflected in the value of E_L? Answering these questions is clearly important in being able to assign R_{aw}, R_t, and E_L their appropriate physiological interpretations. As we shall see shortly, we have answers to some of these questions, but some remain incompletely understood.

4.1   Physics of airway resistance

We all have some experiential equivalent of drinking through straws of different diameters; to get a drink, one has to suck hard on a thin straw and not so hard on a wide one. This experience makes us comfortable with the concept of an airway having a resistance to air flow that reflects its physical dimensions; narrow airways have a large resistance, and vice versa. Such notions are stated precisely in mathematical terms in the equations of fluid dynamics, which trace their beginnings at least as far back as Isaac Newton. Accordingly, our understanding of R_{aw} is well grounded in the basic principles of physics.

Figure 4.1 Laminar flow occurs at low Reynolds numbers, when the flow streamlines are well defined and orderly. Turbulent flow is fully developed at Reynolds numbers well above 2000, when the movement of each parcel of gas is chaotic and the streamlines are mixed up.

4.1.1 Viscosity

When air molecules flow through a conduit, the net movement of each molecule is in the direction of bulk flow. However, not all molecules move in precisely the same direction, or at precisely the same speed, so they periodically bump into each other as they flow along the conduit. These collisions take place randomly, causing the molecules involved to exchange random amounts of kinetic energy and momentum (always preserving the totals of both quantities, of course). The molecules also rebound from each other in random directions. Consequently, as gas flows along the conduit there is a progressive increase in the random motion of its molecules, otherwise known as heat. This heat energy is then passed on to the walls of the conduit and beyond. Molecular kinetic energy in a flowing stream of gas thus gets gradually converted from an organized linear form into a disorganized random form that is eventually dissipated into the environment.

At a more macroscopic level, one can envisage gas flowing along a conduit as occurring in parallel sheets, or cylinders, that move in the direction of flow. The bulk flow of the gas is given by the sum of the flows due to each sheet. In general, however, the gas also experiences *shear*, which means that adjacent sheets move at different velocities. Consequently, as they slide past one another they experience frictional forces due to the molecular interactions described above. The coefficient of this friction is called *viscosity*. A fluid is known as *Newtonian* (i.e. as first described by Isaac Newton) if its viscosity is independent of shear rate. The viscosity of a gas depends on the way in which its molecules interact, and is different for different gases.

4.1.2 Laminar and turbulent flow

The view of gas flow just described, that of thin sheets moving smoothly along in the direction of flow, pertains to a particular type of flow known as *laminar*. When the flow velocity is small and steady, fluid sheets move smoothly along without intermingling to any significant extent. This can be visualized by injecting a thin stream of smoke into the flow. The smoke will show the flow *streamlines* to be distinct and well preserved (Fig. 4.1A). As the flow velocity increases, however, a transition in the nature of the flow starts to occur; the streamlines begin to mix. Eventually, at high enough flows, each little parcel of gas will pursue a chaotic trajectory that includes vigorous lateral movement.

This causes the streamlines to disappear in what is termed fully developed turbulent flow (Fig. 4.1B).

Whether flow through a conduit is laminar, turbulent, or some intermediate combination of both is governed to a large extent by the *Reynolds number*, Re, defined as

$$Re = \frac{2\rho r v}{\mu}$$
$$= \frac{2\rho \dot{V}}{\pi \mu r} \tag{4.1}$$

where ρ is gas density, r is conduit radius, v is linear flow velocity, and μ is gas viscosity. Re is one of the numerous dimensionless numbers used in fluid mechanics to define flow conditions in a manner that is independent of the size of the conduit in question. Theoretical understanding of the factors governing transition from laminar to turbulence is still incomplete, but it has long been established experimentally that flow is laminar when Re is significantly less than about 2000, turbulent when Re is well above 2000, and in transition between these two extremes [3].

The peak flows during resting ventilation in an adult human are of the order of 1 L.s^{-1} in the trachea, which has a radius of about 1 cm. Using values for air of $\rho = 1.2 \times 10^{-3}$ g.mL^{-1} and $\mu = 2 \times 10^{-4}$ g.cm^{-1}.s^{-1} gives an Re of about 4000. During exercise this can increase several-fold, so flow conditions in the trachea are at least bordering on turbulent. The situation is different in the smaller airways. By the time the airway tree has branched half a dozen times, its airways are approaching 1 mm in radius, but the total flow of 1 L.s^{-1} is now divided between the $2^6 = 64$ parallel airways of the sixth generation. This gives a value for Re in each branch of the sixth generation of only about 600, which puts it in the laminar regime.

Re is not the only factor relevant to the nature of flow in an airway. Even when Re is small, it can take some time for the streamlines to settle into their steady-state configuration once flow begins. Flows in the pulmonary airways are continually reversing direction with the phase of the breathing cycle, which raises the question as to whether there is enough time for the streamlines to ever reach a steady state. If the streamlines are fully developed most of the time then flow is called *steady*. Otherwise it is *unsteady*. This issue is determined by the value of another important dimensionless quantity known as the *Womersley number*, α, defined as

$$\alpha = r\sqrt{\frac{2\pi f \rho}{\mu}}, \tag{4.2}$$

where f is respiratory frequency. The transition between steady and unsteady flow occurs when α has a value in the region of 1 [3]. In the airway tree, α achieves its greatest value in the trachea, the airway with the largest r. At a normal breathing frequency of 0.2 Hz, α has a value of about 2.7, so flow conditions in the lung are bordering on unsteady, meaning that streamlines barely have a chance to establish themselves during one phase of the breathing cycle before flow is reversed.

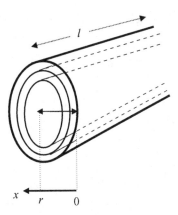

Figure 4.2 Gas flow through a conduit of circular cross-section, radius r, occurs in a series of concentric cylinders. Linear velocity is zero at the conduit wall ($x = 0$) and increases toward the center ($x = r$), where it is maximum.

Entrance effects due to the geometry of the upstream channel exert another important influence on flow streamlines in an airway. If this geometry is different to that of the airway in question, streamlines become fully developed only when the gas has had a chance to flow along the airway for a length equal to about three airway diameters, but by this point the gas is about to enter the next airway generation.

Air flow in the lung is thus neither perfectly steady nor fully developed. Nevertheless, fully developed steady laminar flow conditions are frequently assumed in mathematical models of the lung because they allow calculation of R_{aw} from first principles in a manner that precisely links airway structure to function.

4.1.3   Poiseuille resistance

Consider a gas flowing along a conduit with circular cross-section when the flow is steady, laminar, and fully developed. From the symmetry of the situation, the linear flow velocity (v) is a function only of distance (x) from the inner wall of the conduit to its axial center. The gas flowing along the conduit can thus be thought of as moving along in a set of concentric cylinders (Fig. 4.2). The cylinders will, in general, be moving at different speeds, but the radial distribution of these velocities is not random. Sliding friction between adjacent cylinders tends to keep their velocity differences small in order to minimize the rate of energy dissipation. This also applies to the velocity difference between the outermost cylinder and the adjacent airway wall. In the limit as the thickness of this outer cylinder becomes infinitesimal, its velocity relative to that of the stationary wall also becomes infinitesimal. In other words, the outer cylinder of gas right next to the wall is itself effectively stationary. This is the so-called *no-slip condition* at $x = 0$, where $v(0) = 0$.

The radius of a cylinder located a distance x from the conduit wall is $r - x$. If the wall thickness of this cylinder is dx, then it carries a flow of $2\pi(r - x)v(x)\,dx$. Total flow is the sum of the flows due to all such cylinders, which is

$$\dot{V} = \int_0^r 2\pi \, (r - x) \, v(x) \, dx. \tag{4.3}$$

From the circular symmetry of the conduit we also know that $v(x)$ must be symmetric either side of the axial center. This means that

$$\left. \frac{dv(x)}{dx} \right|_{x=r} = 0. \tag{4.4}$$

Given the chance, dynamic processes in nature always try to organize themselves in the most energy-efficient manner possible. Accordingly, the $v(x)$ that pertains under steady-state conditions is that which requires the minimum power to drive a given bulk flow along the conduit. We can calculate the $v(x)$ that achieves this as follows. The inner of two concentric fluid cylinders moves faster than the outer cylinder by the amount $dv = (dv/dx) \, dx$. This generates a frictional force between the two cylinders that dissipates energy at a rate proportional to both $dv(x)/dx$ and μ. The total rate of energy dissipation, $\dot{W}$, per unit length of conduit is the integral of all these inter-cylinder rates, and is given by

$$\dot{W} = \int_0^r 2\pi \mu (r - x) \frac{dv(x)}{dx} \, dx. \tag{4.5}$$

The question now is to find the functional form for $v(x)$ that minimizes $\dot{W}$ given the constraint that total flow is fixed by Eq. 4.3. To accommodate all possibilities, let $v(x)$ be a general polynomial function of x; that is,

$$v(x) = \sum_{n=1}^{\infty} a_n x^n \tag{4.6}$$

where we note that there is no term containing a_0 because of the no-slip condition at $x = 0$. Substituting Eq. 4.6 into Eqs. 4.3 and 4.5 gives

$$\dot{V} = \int_0^r 2\pi \, (r - x) \sum_{n=1}^{\infty} a_n x^n \, dx$$

$$= 2\pi \sum_{n=1}^{\infty} \frac{a_n r^{n+2}}{(n+1)(n+2)} \tag{4.7}$$

and

$$\dot{W} = \int_0^r 2\pi \mu (r - x) \sum_{n=1}^{\infty} n a_n x^{n-1} \, dx$$

$$= 2\pi \mu \sum_{n=1}^{\infty} \frac{a_n r^{n+1}}{n+1}. \tag{4.8}$$

Minimizing $\dot{W}$ in Eq. 4.8 subject to the constraint given by Eq. 4.7 is achieved by the technique of *Lagrange multipliers*, in which the cost function

$$J = 2\pi \mu \sum_{n=1}^{\infty} \frac{a_n r^{n+1}}{n+1} + \lambda \left[2\pi \sum_{n=1}^{\infty} \frac{a_n r^{n+2}}{(n+1)(n+2)} - \dot{V} \right] \qquad (4.9)$$

is minimized with respect to the a_n and to λ, the so-called Lagrange multiplier. Minimization is achieved by differentiating J with respect to each parameter in turn and setting the result equal to zero. In particular, this gives

$$\frac{\partial J}{\partial a_n} = \frac{2\pi r^{n+1}}{n+1} \left[\mu + \lambda \frac{r}{(n+2)} \right]$$
$$= 0, \qquad (4.10)$$

from which we obtain

$$\lambda = \frac{-(n+2)\mu}{r}. \qquad (4.11)$$

Now, Eq. 4.11 implies that λ takes a different value for every value of n. This means that λ can only have a well-defined optimum value that minimizes Eq. 4.9 if there is only a single independent value of a_n to be determined, namely a_1. The value of a_2 is determined by a_1 because, according to Eq. 4.4,

$$\frac{d}{dx} \left[a_1 x + a_2 x^2 \right]_{x=r} = a_1 + 2a_2 r$$
$$= 0 \qquad (4.12)$$

so that

$$a_2 = \frac{-a_1}{2r}. \qquad (4.13)$$

There can be no other terms, meaning that

$$v(x) = a_1 x - \frac{a_1}{2r} x^2. \qquad (4.14)$$

In other words, $v(x)$ is a parabolic function of x. Substituting Eq. 4.14 into Eq. 4.7 and rearranging gives

$$a_1 = \frac{4\dot{V}}{\pi r^3}. \qquad (4.15)$$

To calculate the driving pressure that this flow regime requires, consider the balance of forces acting on one of the fluid cylinders depicted in Fig. 4.2. Because there is no acceleration, the force acting on the upstream face of the cylinder is balanced by the net frictional forces acting between the cylinder and its two adjacent neighbors. The force acting on the cylinder face is due to P acting over the face area of $2\pi(r - x)dx$. The inner of the neighboring cylinders is moving faster than the cylinder in question, while the outer cylinder is moving slower. The cylinder is thus pulled along from the inside (a negative contribution to the force balance) and slowed down from the outside (a positive contribution). The force balance equation is therefore

$$
\begin{aligned}
2\pi(r-x)P\,dx &= 2\pi(r-x)\mu l \frac{dv}{dx}\bigg|_x - 2\pi(r-x-dx)\mu l \frac{dv}{dx}\bigg|_{x+dx} \\
&= 2\pi(r-x)\mu l \frac{dv}{dx}\bigg|_x - 2\pi(r-x-dx)\mu l \left(\frac{dv}{dx}\bigg|_x + \frac{d^2v}{dx^2}\bigg|_x dx \right) \\
&= 2\pi\mu l \left[\frac{dv}{dx}dx - (r-x)\frac{d^2v}{dx^2}dx \right]
\end{aligned}
\tag{4.16}
$$

where the second line of the above equation was derived from the first line using a first-order Taylor series expansion, terms containing squared differentials are taken as negligible, and l is the length of the conduit. Using Eqs. 4.13–4.15 gives

$$
\begin{aligned}
P &= \mu l \left[\frac{1}{(r-x)}\frac{dv}{dx} - \frac{d^2v}{dx^2} \right] \\
&= \mu l \left[\frac{a_1(1-x/r)}{r-x} + \frac{a_1}{r} \right] \\
&= \frac{8\mu l \dot{V}}{\pi r^4}.
\end{aligned}
\tag{4.17}
$$

In other words,

$$
\begin{aligned}
R &= \frac{P}{\dot{V}} \\
&= \frac{8\mu l}{\pi r^4}.
\end{aligned}
\tag{4.18}
$$

Equation 4.18 is the well-known formula for the Poiseuille flow resistance of a circular conduit that often appears in texts on pulmonary physiology [39, 40]. The parabolic velocity profile of Poiseuille flow arises because it is theoretically the most energy efficient. Nevertheless, as the above derivation shows, this formula is based on a number of specific and rather constraining assumptions. These assumptions are likely to be violated by a significant margin in practice, even in the airways of a normal lung during quiet breathing. Consequently, the actual resistance of an airway branch will invariably be somewhat larger than that given by the Poiseuille formula, particularly the branches of the first few generations. On the other hand, the inverse dependence of R on the fourth power of r remains a useful description of reality. It even applies reasonably well if flow is turbulent, although in this case R depends roughly linearly on $\dot{V}$ rather than being independent of it [3]. By contrast, R depends only linearly on l, so a change in airway caliber is a much more powerful modulator of R than is a change in airway length.

4.1.4   Resistance of the airway tree

Once the resistance of each branch in the airway tree has been determined, R_{aw} itself can be obtained by adding the individual resistances in series or parallel, as the case may be. The building block for this process consists of a single parent airway, resistance R_p, together with its two daughter airways having resistances R_{d1} and R_{d2}, respectively (Fig. 4.3). The pressure drop across each of the two daughters, ΔP_d, is the same but they

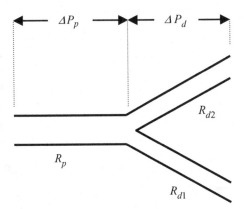

Figure 4.3 A parent airway with resistance R_p connects to two daughter airways with resistances R_{d1} and R_{d2}. The pressure across the parent is ΔP_p and the pressure across each daughter is ΔP_d.

share the total flow, $\dot{V}$, so

$$\dot{V} = \frac{\Delta P_d}{R_{d1}} + \frac{\Delta P_d}{R_{d2}}. \tag{4.19}$$

This gives the combined resistance of the two daughters, R_d, as

$$R_d = \frac{\Delta P_d}{\dot{V}}$$

$$= \frac{R_{d1} R_{d2}}{R_{d1} + R_{d2}}. \tag{4.20}$$

By contrast, the parent carries the same flow as the two daughters together, but the pressure drop across the parent, ΔP_p, adds to ΔP_d to give the total pressure drop, ΔP, across the entire assembly. That is,

$$\Delta P = \Delta P_p + \Delta P_d$$

$$= \dot{V}(R_p + R_d). \tag{4.21}$$

The total resistance, R_{tot}, is the combined resistances of the parent in series with two parallel daughters, and is given by

$$R_{tot} = \frac{\Delta P}{\dot{V}}$$

$$= R_p + \frac{R_{d1} R_{d2}}{R_{d1} + R_{d2}}. \tag{4.22}$$

The formula in Eq. 4.22 can be used recursively to make calculations of R_{aw} for the entire airway trees of various species, including humans, based on information about the diameters and lengths of the branches at each generation. The most simplistic calculations assume Poiseuille flow in each airway. Better approximations to reality are obtained by applying empirical correction factors to account for the effects of turbulence, upstream geometry and unsteadiness. This is particularly important for the more proximal airway generations where Re is highest [41]. Currently, the most advanced calculations of R_{aw} employ numerical methods to solve the Navier–Stokes

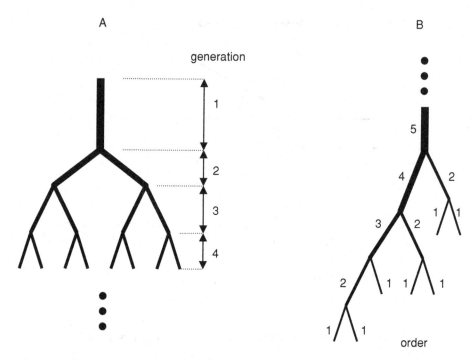

Figure 4.4 (A) The symmetrically bifurcating airway tree structure in which airways are specified by generation number, beginning at the trachea. (B) The asymmetrical self-similar tree structure in which airways are specified by order number, beginning with the terminal bronchioles.

equations governing fluid flow in anatomically accurate three-dimensional models of the airway tree [42]. Such methods are computationally very expensive, but provide an extremely detailed link between airway structure and function.

The branching structure of the airway tree strongly influences the value of R_{aw}. The simplest structure is the dichotomously branching scheme (Fig. 4.4A) in which each parent airway gives rise to two identical daughters, and all the branches of a given *generation* are identical [43], where the trachea is generation 1, the main stem bronchi comprise generation 2, and so on. The number of airways in generation n is thus 2^{n-1}. Modeling this kind of tree is particularly convenient because R_{aw} is simply the sum of the resistances of each generation. Furthermore, the combined resistance of all the airways in generation n is given by the resistance of one of the airways divided by 2^{n-1}. Figure 4.5 shows a prediction of the distribution of resistance with generation number down the airway tree for a human for various bulk flow rates [41], and shows that most of the contribution to R_{aw} comes from the proximal airways. In other words, although the distal airways become progressively smaller, this is more than offset by the geometric increase in their number.

A more realistic representation of the airway tree acknowledges that it branches asymmetrically. A computationally convenient way to codify this asymmetry is to assume that the tree is self-similar, which means that the degree of asymmetry at any point in the tree is the same as at all other points at the same level [44]. To do this it is necessary to define the "level in the tree" in terms of *airway order* rather than generation. Order

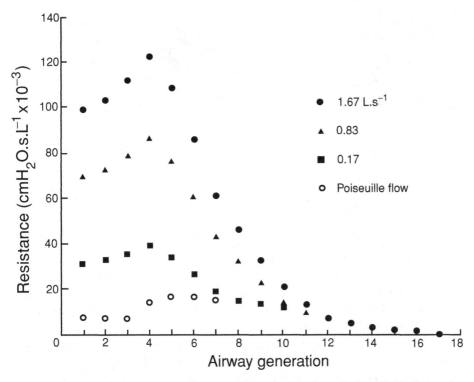

Figure 4.5 Resistance as a function of airway generation calculated for the human lung based on a bifurcating model of the airway tree (reproduced with permission from [41]).

numbering begins at the distal end of the tree rather than at the trachea. Thus, all terminal bronchioles are of order 1. However, not all parents of order 1 airways have to be of order 2. Some parents may be of order 3 or higher. An airway of order n still branches into daughter airways, but one is of order $n - 1$ and one is of order $n - 1 - \delta$. The larger the value of δ, the larger the degree of asymmetry in the tree downstream of the branch point. The tree becomes self-similar if δ is a function only of airway order; in other words, exactly the same tree structure is subtended by all airways of a given order (Fig. 4.4B). To completely specify a self-similar airway tree, one merely needs to list the airway dimensions (radius and length) and δ value for each order in the tree [44–47]. This is much more efficient than having to list the dimensions of every branch in the tree.

The human airway tree is reasonably symmetric, with δ taking values between 0 and 3 [48]. Many animal species have airway trees with much greater asymmetry. The mouse, for example, achieves δ values of up to 6 [49].

4.2   Tissue resistance

As we established in Section 3.1.2, the resistance of the lung (R_L) is due not only to the flow resistance of the airways. R_L also contains a significant contribution from

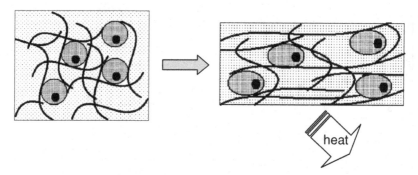

Figure 4.6 When lung tissue is stretched and deformed (left to right), its constituent cells, fibers and fluids slide past each other and generate heat that is dissipated into the surroundings.

the lung tissues (R_t). Furthermore, the dissipative properties of lung tissue can change substantially in lung diseases such as asthma [50] and pulmonary fibrosis [51]. Linking R_t to tissue structure, however, is not nearly as straightforward as linking R_{aw} to airway structure. This is because lung tissue is a much more complicated material than air. Nevertheless, the basic mechanism giving rise to R_{aw} and R_t is essentially the same. This mechanism is internal friction. When lung tissue is stretched, as occurs during changes in lung volume, its constituent fibers, cells, and fluids move against each other. This produces friction and heat (Fig. 4.6), just as occurs in a flowing stream of gas due to viscosity, even though the details of the events occurring with tissue are much more complicated.

Thus, while we have a very good quantitative understanding of the physical processes that link the geometry of a conduit to R_{aw}, our understanding of how R_t arises remains very much at the qualitative level. Nevertheless, we do have a great deal of empirical information about how lung tissue behaves mechanically thanks to the alveolar capsule technique, which allows R_t to be measured directly (Eq. 3.9). We know, for example, that R_t increases during bronchoconstriction, which occurs when a smooth muscle agonist is administered to the airways. This causes the smooth muscle surrounding the airways to contract and narrow the airway lumen [52, 53]. Airway narrowing in turn causes distortion of the parenchymal tissues within which the airways are embedded. This can change the rheological properties of the tissues [54, 55]. Similarly, R_t increases with transpulmonary pressure [56] because of the stress that is applied to the tissues. Pathological conditions involving alterations in tissue structure and composition, such as fibrosis [57, 58] and emphysema [59, 60], can also affect R_t.

4.3   Lung elastance

4.3.1   The effect of lung size

The value of the parameter E in the model in Fig. 3.4, like R_t, is a reflection of the rheological properties of the lung tissues. E is a measure of how difficult it is to stretch this tissue in order to increase the volume it surrounds, so obviously its value must have something to do with what the tissue is made of. However, the identities of the tissue constituents and their relative abundances are not the only factors that determine E; probably more important is how the constituents are arranged with respect to each other. Nevertheless, the intrinsic nature of lung tissue does not vary much between animals of different sizes, so the primary determinant of E is lung size. For example, if one was to suddenly halve the size of the lung then its remaining tissues would have to stretch twice as much to accommodate a given volume of air. The resulting elastic recoil pressure in the lung would increase accordingly, and so consequently would E. In fact, the same argument also applies to R_t.

The lung volume effect explains why E changes in pathological situations that cause some fraction of the lung to become blocked from receiving ventilation. This process is known as *derecruitment* and may occur at the level of the alveoli or the airways. Derecruitment of alveoli can occur, for example, with pulmonary edema when fluid leaks into the distal airspaces of the lung from the vasculature and displaces all the air [61]. Alveolar derecruitment can also occur when pulmonary surfactant does not work as it should to mitigate the forces of surface tension. This can result in alveoli collapse, a phenomenon known as *atelectasis*. Derecruitment of small airways in the lung also tends to occur with surfactant dysfunction, increased airway secretions, or airway narrowing. These factors can cause liquid bridges to form across the airway lumen [62], closing off all downstream regions of the lung. Derecruitment is a typical feature of acute lung injury [61, 63–65] and attacks of asthma [66], and causes E to increase in inverse proportion to the fraction of lung volume that has been lost.

Finally, E will change in pathological situations that lead to a change in the intrinsic stiffness of the lung tissue. Tissue stiffness is increased, for example, in pulmonary fibrosis because of the abnormal levels and arrangement of connective tissue proteins that accompany this disease [57]. Conversely, stiffness is decreased in emphysema as a result of the destruction of alveolar walls that are responsible for transmitting elastic forces across the lung [67]. In clinical applications, E is usually expressed as its inverse, *pulmonary compliance*.

4.3.2 Surface tension

The foregoing might suggest that one should think of lung tissue primarily as a kind of naturally elastic material like rubber. However, the most important factor responsible for the elastic recoil of an intact lung arises not from the tissue components themselves but rather from the surface of the liquid that lines all the airspaces of the lung, the alveoli in particular. Water molecules are polar, which causes them to want to align themselves so that positively and negatively charged regions of adjacent molecules are juxtaposed. This gives rise to a horizontally directed force holding the molecules together at the surface, known as *surface tension*. The extent of the surface tension effect in the lungs is dramatically demonstrated by observing how much easier it is to inflate lungs when they are filled with saline and the air-liquid interface is no longer present (Fig. 4.7).

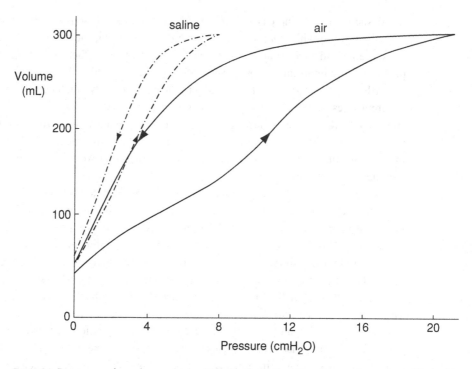

Figure 4.7 Pressure–volume loops obtained by slowly cycling the volume of a cat lung filled with either air or saline (adapted with permission from [68]).

The importance of surface tension for the lung arises from the fact that that an alveolus is a very delicate membranous sac lined with liquid, and so acts somewhat like a bubble of fluid with an air-liquid interface at its inner surface (Fig. 4.8). The surface tension on the inner surface of the bubble is T with units of force per unit length, which means there has to be a positive pressure, P, inside the bubble to prevent it from imploding.

We can relate T and P by considering a virtual surface passing through the equator of the bubble (Fig. 4.8). The net force pushing down on the part of this surface that is inside the bubble is $\pi r^2 P$, so this must equal the reactive force pushing upward on the upper half-dome of the bubble. This upward force on the upper dome is balanced by the surface tension forces pulling down on its equatorial perimeter, and equals $2\pi r T$. Equating $\pi r^2 P$ and $2\pi r T$ gives

$$P = \frac{2T}{r}. \tag{4.23}$$

Equation 4.23 is known as *Laplace's law*, and shows that for a given value of T, the pressure inside a liquid bubble increases as the radius of curvature of the bubble decreases. The importance of Laplace's law for the lung arises from the fact that the alveoli are extremely small, so r in Eq. 4.23 is also small. This causes P to be correspondingly large. For example, the surface tension of pure water is about 7.2×10^{-4} N.cm^{-1}, while the radius of curvature of an alveolus is of the order of 10^{-2} cm. Using these values in

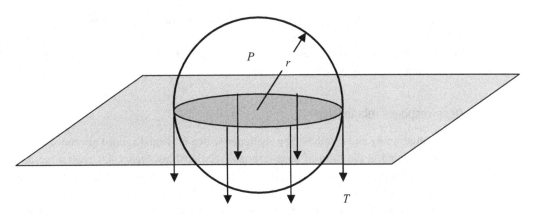

Figure 4.8 A spherical bubble of liquid with surface tension T and radius r maintains an internal pressure of P.

Eq. 4.23 we find that P inside a bubble of water the size of an alveolus is about 1.5×10^{-1} N.cm^{-2}, which is about 15 cmH$_2$O in the usual units of air pressure employed by pulmonary physiologists. In other words, if the alveolar lining fluid were pure water we would need to maintain a baseline pressure across the lung tissue of 15 cmH$_2$O just to keep the alveoli open, let alone inflate them further in order to take a breath. This pressure is much larger than that normally required to maintain the lung at end-expiratory volume. Of course, the alveoli are not perfectly spherical, nor are they mechanically independent of the structures around them, so the above calculations should be taken as a conceptual illustration of the problem caused by surface tension rather than an accurate assessment of intrapulmonary pressures [69].

Fortunately for us and our fellow mammals, nature has come up with a solution to the potentially lethal problem of surface tension in the form of pulmonary surfactant, a large bipolar molecule rather like detergent that is secreted by the type II epithelial cells that form part of the alveolar lining. Surfactant lowers the surface tension of the air-liquid interface within the lung to the point that only 2 or 3 cmH$_2$O is required to inflate the lungs to functional residual capacity. Surfactant also modifies the surface tension of the air-liquid interface in a way that varies with how much an alveolus is inflated, so that two communicating alveoli of different sizes can co-exist stably. If the alveoli were lined with only water, the smaller alveolus would have the greater pressure and would tend to collapse into the larger one. The crucial importance of surfactant in reducing E can be demonstrated by performing a saline lavage of the lung. This washes out much of the surfactant at the air–liquid interface and causes an immediate increase in lung stiffness that can rise to several times normal values [70].

4.4 Resistance and elastance during bronchoconstriction

One of the ways in which R_L and E_L are most widely used is to quantify the response of the lungs to bronchial challenge. When a smooth muscle agonist such as methacholine is administered to the lungs, either as an inhaled aerosol or an intravenous injection, the lungs undergo a transient process known as bronchoconstriction. The main event taking place during bronchoconstriction is narrowing of the conducting airways due to contraction of smooth muscle in the airway wall.

4.4.1   Dose-response relationship

The changes in R_L and E_L elicited by challenge with a bronchial agonist are conventionally represented in the form of dose-response relationships. Figure 4.9 gives a typical example in mice instrumented with an alveolar capsule, and shows how the components of R_L (R_{aw} and R_t) and E_L increase from baseline through a series of doubling doses of methacholine [71]. Dose-response curves such as those in Fig. 4.9 are characterized by two key attributes: (1) the dose of agonist required to elicit a response significantly above baseline, which gives a measure of *sensitivity*, and (2) the response elicited by the maximum dose of agonist, which reflects *responsiveness*. Note also that allergically inflamed mice (closed circles in Fig. 4.9) are both hypersensitive and hyperresponsive to methacholine, which are characteristics of asthma. Furthermore, these effects are observed to varying degrees in all three mechanical parameters, R_{aw}, R_t, and E_L.

It is easy to see why bronchoconstriction causes R_{aw} to increase; contraction of airway smooth muscle narrows the airway lumen and thus increases its resistance to flow [3, 72]. Somewhat less obvious is why bronchoconstriction should also increase R_t, but this can be explained in part noting that the airways are intimately embedded within and attached to the lung parenchyma. When the airways narrow they induce local distortion of the tissue, which places it under increased stress [55]. Consequently, more frictional heat is generated within the tissue as it is stretched during lung inflation. R_t can also appear to increase when the lung behaves more like a number of different airway-tissue compartments instead of the single-compartment model shown in Fig. 3.4 [46, 73]. This multi-compartment behavior is something that always seems to accompany the development of bronchoconstriction [74], and will be treated in detail in Chapter 7. Suffice it to say at this point that bronchoconstriction causes both R_{aw} and R_t to increase commensurately with the degree of smooth muscle contraction taking place in the airway walls.

Another aspect of the data shown in Fig. 4.9 that might seem curious is that E_L increased during bronchoconstriction at least as much as either R_{aw} or R_t. In fact, this is typical; measurements R_L and E_L are invariably tightly linked and almost never change in isolation. Much of the reason for this coupling between E_L and R_L stems from the heterogeneous nature of the lung's response to bronchial agonists, which will be covered in later chapters. Even in a homogeneous lung, however, one would expect a change in airway caliber to affect E_L for the same reason that it affects R_t; most of the airways are embedded within the lung parenchyma and attached to it [55]. Consequently, when an airway narrows in response to a bronchial agonist, the local distortions induced in the adjacent tissue increase the stiffness of the parenchyma. This *airway-parenchymal interdependence* works the other way, too. Inflation of the parenchyma increases tension in the tissue attached to the outside of the airway wall, leading to an increase in the

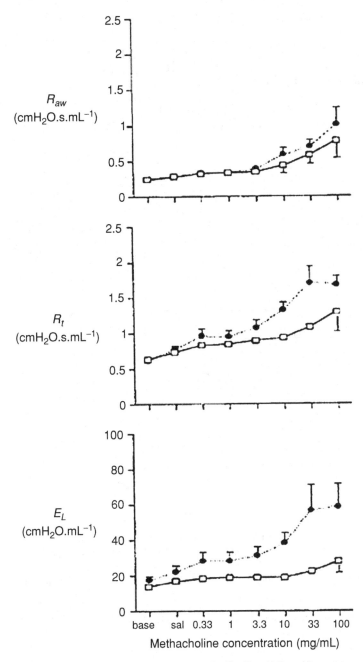

Figure 4.9 Dose-response curves to aerosols of saline (sal) and increasing concentrations of methacholine for R_{aw}, R_t, and E_L. The open circles were obtained from normal BALB/c mice. The closed circles were obtained from BALB/c mice that were previously sensitized and challenged with ovalbumin to produce an inflammatory response in the lungs similar to that often seen in asthma (adapted with permission from [16]).

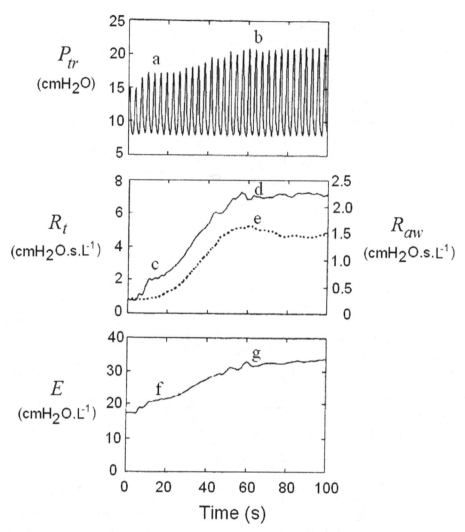

Figure 4.10 A bolus injection of histamine was given into the vena cava of a dog through a femoral catheter at time = 0 s. The subsequent response in P_{tr} exhibits two phases labeled a and b. These phases are mirrored in R_L (c and d, respectively) and in E_L (f and g, respectively). By contrast, R_{aw} exhibits only the second phase of the response (e) (reproduced with permission from [71]).

caliber of the airway lumen and giving R_{aw} an inverse dependence on lung volume [52, 75].

4.4.2 Time-course of bronchoconstriction

Dose-response relationships for resistance and elastance consider these parameters as purely static functions of the level of an intervention. Figure 4.10 shows examples of R_L and E_L obtained in a dog for the 100 s immediately following an intravenous injection of the bronchial agonist histamine [71]. The P_{tr} signal recorded while the animal was mechanically ventilated with a fixed tidal volume over this time period shows a biphasic response that is reflected in both R_L and E_L. However, the continuous tracing of R_{aw} obtained from simultaneous alveolar capsule measurements of P_A via Eq. 3.8 exhibits only the second phase of the response. These data are interpreted as reflecting the dynamics of delivery of histamine to the airway smooth muscle. The drug was injected into the vena cava so it first reached the periphery of the lung via the pulmonary circulation where it was able to activate the smooth muscle in only the small peripheral airways and alveolar ducts. This gave rise to the first phase of the response, which was manifest only in the parameters pertaining to the mechanical properties of the tissues, namely E_L and the tissue component of R_L, but not R_{aw}. Subsequently, the histamine moved on to the systemic circulation, which includes the bronchial supply, where it was able to reach the majority of the airway smooth muscle distributed throughout the airway tree. This caused the response in R_{aw}.

Interestingly, the corresponding second phase responses in R_L and E_L were substantially greater than their initial responses, which would seem to indicate that a constriction of the central airways affects the apparent mechanical properties of the lung tissues. This could have been due in part to the airway-parenchymal interdependence effects described above. Most of it, however, was probably due to the fact that airway narrowing occurs heterogeneously throughout the lung, as will be discussed later. In any case, this example shows that a study of the temporal dynamics of bronchoconstriction reveals physiological insights that are not available from the more conventional dose-response relationship. Tracking model parameter values as they change in time can thus be a powerful investigative method.

4.4.3 Determinants of airways responsiveness

The ultimate goal of measuring R_L or E_L in patients or animal models of lung disease, is to link the structure of the lung to its function. In the case of airway resistance, the link is obvious; an increase in R_{aw} means that the airways have narrowed. Subjects differ, however, in the amount by which their R_{aw} will increase when administered a standard dose of an agonist like methacholine. When the increase is excessive, the subject is said to be *hyperresponsive*. Patients with asthma are hyperresponsive compared to normal individuals. Allergically inflamed mice are hyperresponsive compared to control animals (Fig. 4.9). Measuring lung mechanics is thus crucial for the investigation of airways hyperresponsiveness.

Of course, once a lung has been found to be hyperresponsive, the next step is to determine the underlying cause. The primary effector of bronchoconstriction is the airway smooth muscle, which is stimulated to contract by the administered agonist. Because this muscle is wrapped around the airway, when it shortens it compresses the airway walls inward, thereby narrowing the lumen. Consequently, the explanation for hyperresponsiveness that most readily springs to mind is that the airway smooth muscle must be stronger than normal. In fact, the situation is considerably more complicated than this [76, 77], which again illustrates the importance of not jumping to conclusions. For example, smooth muscle can only narrow the lumen of the airway it surrounds if it

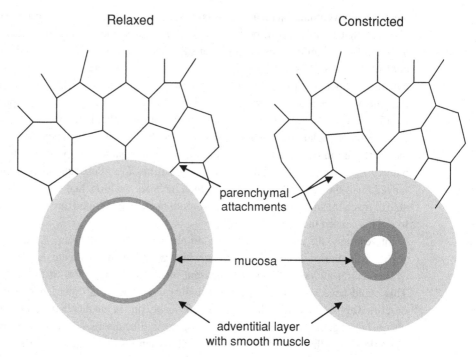

Relaxed   Constricted

parenchymal
attachments

mucosa

adventitial layer
with smooth muscle

Figure 4.11 A relaxed airway in cross-section, embedded in parenchyma, is shown on the left. The light gray annulus represents the adventitial tissue of the airway wall together with the airway smooth muscle. The dark gray annulus represents the inner mucosal layer that abuts onto the airway lumen itself (shown in white). When the smooth muscle contracts, the adventitial and mucosal volumes are preserved, so they thicken, as shown on the right. Also, the parenchyma in the region of the airway becomes distorted as it is pulled inward.

can generate enough force to overcome the forces opposing it. The principal opposition is provided by the elastic recoil of the airway wall and the outward tethering forces exerted by the parenchyma within which the airway is embedded (Fig. 4.11). The chronic inflammation associated with diseases such as asthma has been postulated to cause the airway wall to remodel, which could affect its stiffness and hence the balance of forces that determine how much airway narrowing occurs when the smooth muscle is activated. It remains controversial as to whether remodeling is likely to enhance or reduce airways responsiveness [78].

The outward tethering forces exerted by the parenchymal attachments to the outside of the airway wall have an enormous influence on the ability of the smooth muscle to shorten. The attachments transmit transpulmonary pressure across the airway wall, and it is this pressure that opposes shortening of the airway smooth muscle. Consequently, increasing lung volume causes a dramatic decrease in airways responsiveness [75, 79].

Airways responsiveness is not only determined by a balance of forces. Geometric factors can also have a major influence. For example, when the airway smooth muscle shortens around the airway, it compresses the tissue making up the inner part of the airway wall, notably the mucosal lining. The volume of this lining is thought to be preserved during narrowing, so it necessarily becomes thicker as shown in Fig. 4.11. This causes the lumen to be reduced much more rapidly than would be the case if no mucosal lining were present [80]. Accordingly, any situation in which the mucosa becomes thickened, such as during the inflammation caused by an allergic reaction, can enhance luminal narrowing even when the strength of the smooth muscle and the elastic forces opposing it are completely normal. Accumulations of airway secretions can have a similar effect [81].

These and other factors individually have the potential to affect airways responsiveness to a substantial degree [76, 82], so it is tempting to speculate about which might be the most potent and likely to be responsible for a disease such as asthma. In reality, these factors probably co-exist. Indeed, they can act together in a synergistic fashion [83], so it is likely that multiple factors are at play.

Problems

4.1 Calculate the Reynolds number (Re) for a circular pipe of radius 1 cm through which air flows at a steady rate of 1 L.s^{-1} (the density of air is 1.3×10^{-3} g.mL^{-1}). What is Re if water flows through the conduit at the same rate? How fast do both air and water have to flow through the pipe in order for flow to become turbulent? Repeat the calculation for helium gas.

4.2 Suppose that fluid flows at a constant rate between two flat parallel plates separated by a distance d. Flow is fully developed and laminar, and the plates are much wider than d so that edge effects can be ignored. Assuming no-slip conditions at the surface of the plates, calculate the resistance to flow per unit width and per unit length of the plates.

4.3 Consider a parent airway and its two daughters, as shown in Fig. 4.3. Suppose that the combined cross-sectional area of the two daughters is fixed at A, and that the downstream pressure presented to both daughters is atmospheric. Write down an expression for the total resistance of the three airways. What is the ratio of the areas of the two daughter airways that minimizes total resistance?

4.4 Write a recursive algorithm for calculating the resistance of a self-similar symmetric airway tree obeying the morphology shown in Fig. 4.4B.

4.5 From Fig. 4.7, estimate the factor by which lung elastance during inspiration decreases when the air-liquid interface is obliterated by filling the lungs with saline. What is the factor by which elastance is reduced during expiration? Explain why these two factors are not the same.

4.6 Derive Laplace's law (the equivalent of Eq. 4.22) for a cylindrical airway of radius r lined with a fluid of surface tension T.

4.7 How would the various phases labeled a to g in Fig. 4.10 be affected if histamine were injected into the arterial circulation instead of the venous supply? Check your reasoning by examining the experimental results in [71].

5 Nonlinear single-compartment models

The linear single-compartment model shown in Fig. 3.1 accurately mimics the mechanical behavior of a healthy lung during resting breathing or mechanical ventilation. This is rather remarkable given the lung's enormous structural complexity, and shows that its thousands of airways and millions of alveoli normally act together in a highly coordinated fashion. This ceases to be the case, however, when the lung is forced to operate at volumes or frequencies outside the range of normal breathing, or when pathologies set in. Such conditions degrade the fitting performance of the linear single-compartment model, producing elevated values of SSR (Eq. 3.11) and decreased values of CD (Eq. 3.31). Contrast, for example, the two model fits shown in Fig. 5.1. The data from Patient A shown in the upper panel are the same as shown in Fig. 3.6B, and the fit provided by Eq. 3.5 is excellent with $CD = 0.995$. The data from Patient B in the lower panel are also the ensemble average of 60 s of regular mechanical ventilation, and the fit is still very good with $CD = 0.944$. Nevertheless, there are sizeable deviations between the data and the model fit for patient B, suggesting that something was going on in the lungs of this patient that is not represented in the linear single-compartment model.

To achieve a better fit to the data from Patient B, the model has to be extended by the addition of extra features. This gives it increased freedom to follow the bends and twists in the data. However, it raises a new problem: what are these extra features to be? There are many different yet plausible ways in which the linear single-compartment model can be made more realistic, and more complicated. How do we know when we have added enough extra features? What do we do if each new model mimics a given data set with equal precision? How do we decide which model is best?

Answering these questions is a central theme of this book, and requires an organized and hierarchical approach as pointed out in Chapter 1. We begin the process in this chapter by considering how the linear single-compartment model can be taken to the next level of complexity through the incorporation of nonlinear constitutive properties.

5.1 Flow-dependent resistance

The single-compartment linear model is *linear* because its independent variables (V and $\dot{V}$) are linearly related to the dependent variable P. That is, if you just varied one of these independent variables and kept everything else constant, you would get a linear

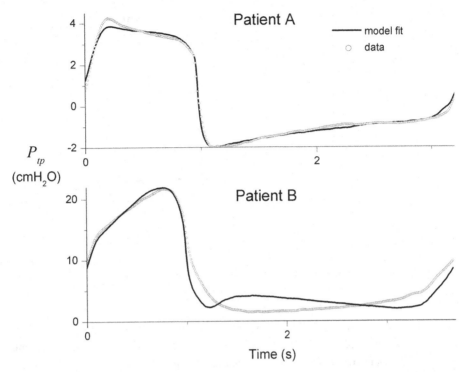

Figure 5.1 Examples of P_{tp} over a breath from two patients (both are ensemble averages of the individual breaths contained in 60 s of data) together with their respective fits provided by the linear single-compartment model.

relationship between P and the variable in question. If the model were to contain a variable to which this did not apply, then the model would be *nonlinear* in that variable. In general, if a model contains any such nonlinear variables, we say the model itself is nonlinear.

As a simple example, suppose that the resistive pressure drop across the airway in the model in Fig. 3.1 were to increase with flow according to the equation

$$\Delta P = K_1 \dot{V} + K_2 \dot{V} |\dot{V}| \tag{5.1}$$

where the vertical bars indicate the absolute value of the enclosed quantity. If Eq. 5.1 is substituted for $R_L \dot{V}$ in Eq. 3.5, the result is

$$P_{tp} = E_L V + K_1 \dot{V} + K_2 \dot{V} |\dot{V}| + P_0. \tag{5.2}$$

This is a nonlinear model because the third term in Eq. 5.2 is nonlinear in the variable $\dot{V}$. However, Eq. 5.2 remains linear in its four parameters, E_L, K_1, K_2, and P_0, which means these parameters can still be evaluated from experimental measurements of P_{tp}, V, and $\dot{V}$ by using multiple linear regression.

In fact, Eq. 5.1 is well known to respiratory physiologists as *Rohrer's equation*, first postulated early in the twentieth century [72]. Rohrer's equation has been shown to apply

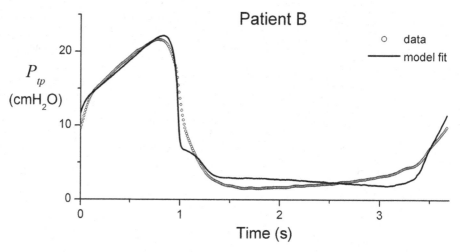

Figure 5.2 Fit of Eq. 5.2 to the P_{tp} data shown in the lower panel of Fig. 5.1. Compare this fit to that in the lower panel of Fig. 5.1.

at high Reynolds numbers when flow is turbulent. It is useful, for example, for describing the pressure-flow relationships of endotracheal tubes at the flows encountered during mechanical ventilation [84]. As pointed out in Chapter 4, turbulence is also expected in the central airways during exercise, or when air follows a tortuous pathway through the larynx.

The above considerations do not guarantee that Eq. 5.2 should replace Eq. 3.5 as a model of the lung. Even if turbulent flow is present in the airways, this does not automatically mean that its effects will be evident in the global pressure-flow behavior of the whole lung. It may be, for example, that the contribution of airway resistance to R_L at normal breathing frequencies is not as important as the contribution of tissue resistance, which, as we shall see in later chapters, has a complicated dependence on flow, tidal volume, and mean lung volume. The nature of the apparent nonlinearity in R_L thus depends very much on the nature of the ventilation pattern that is used during data collection. As shown in Fig. 5.2, the fit provided by Eq. 5.2 to the data in the lower panel in Fig. 5.1 is visually improved to a slight degree over that of the linear model, with CD increasing from 0.944 to 0.972. Nevertheless, the fit from the nonlinear model still shows substantial and systematic deviations either side of the measured P_{tp} signal, suggesting that there is more that needs to be added to the model.

A potential source of confusion arising from Eq. 5.2 concerns the interpretation of the parameters K_1 and K_2. It might seem natural to think that, because we gave a physiological interpretation to their predecessor R_L (Chapter 4), these two parameters should also "mean" something. In fact, K_1 and K_2 cannot be assigned meaningful independent interpretations. There have been attempts to attribute them to the laminar and turbulent components, respectively, of flow, but this has no basis in physical theory [72]. K_1 and K_2 should be considered simply as a pair of parameters that collectively describe a quadratic dependence of resistive pressure on flow. In other words, when

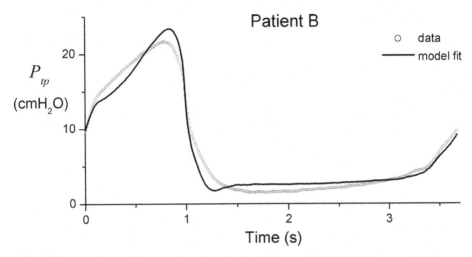

Figure 5.3 Fit of Eq. 5.4 to the P_{tp} data shown in the lower panel of Fig. 5.1. Compare this fit to that in the lower panel of Fig. 5.1 and to the fit in Fig. 5.2.

Rohrer's equation pertains, it no longer makes sense to speak of the airways as having a single fixed resistance. Instead, the airways have a quadratic pressure-flow characteristic given by Eq. 5.1.

5.2 Volume-dependent elastance

5.2.1 Nonlinear pressure-volume relationships

Another possible way to make the single-compartment model nonlinear is to make E_L increase with V. This represents a progressive stiffening of the lung as it inflates from functional residual capacity up to total lung capacity, as is observed experimentally. A simple way to implement this notion is to set

$$P_{el} = E_1 V + E_2 V^2 \tag{5.3}$$

where E_1 and E_2 are constants. The equation of the lung model is then

$$P_{tp} = R_L \dot{V} + E_1 V + E_2 V^2 + P_0, \tag{5.4}$$

which is nonlinear because of the dependence of the third term on V^2. Again, however, this model is linear in its parameters R_L, E_1, E_2, and P_0, so these parameters can still be evaluated from any given data set by multiple linear regression.

Figure 5.3 shows the fit provided by Eq. 5.4 to the data from Patient B. Again, the fit is slightly improved over that of the linear model, with a CD this time of 0.977. Again, however, there is clear room for further improvement, so neither the resistance that depends linearly on flow nor the elastance that depends linearly on volume is sufficient to account for all the deterministic variation in the measured P_{tp} signal. For some data sets, however, Eq. 5.4 does provide a useful improvement in fit over that of the linear

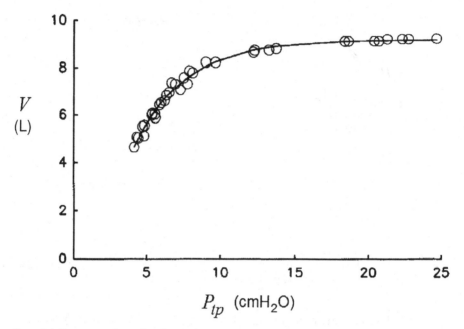

Figure 5.4 Pressure-volume data from a human subject (circles) and the fit provided by Eq. 5.5 (reproduced with permission from [88]).

model, particularly when the lungs are ventilated over a large volume range so that the tissues are stretched toward their nonlinear elastic domain [85] or in pathological situations where lung elasticity is abnormally high [86].

The two nonlinear lung models we have considered so far (Eqs. 5.2 and 5.4) can be fit to experimental data using the technique of multiple linear regression (Section 3.2), which is computationally convenient. Not all nonlinear models of lung mechanics enjoy this particular advantage, however. For example, the quasi-static pressure-volume curve of the lung over the volume range between functional residual capacity and total lung capacity is commonly described by the exponential function [87]

$$V = A - Be^{-KP_{tp}} \qquad (5.5)$$

where A, B, and K are the parameters of the curve. A and B can be obtained by multiple linear regression, but K cannot because it is nonlinearly related to the dependent variable P_{tp}. Accordingly, an iterative model fitting scheme must be used to fit Eq. 5.5 to experimental measurements of P_{tp} and V (Fig. 5.4).

Below functional residual capacity, the quasi-static pressure-volume curve of the lung exhibits an upward concavity (Fig. 5.5). The entire pressure-volume curve of the lung, from residual volume to total lung capacity, thus requires description by a sigmoidal function such as [89]

$$V = a + \left[\frac{b}{1 + e^{-(P_{tp}-c)/d}} \right] \qquad (5.6)$$

where a, b, c, and d are adjustable parameters.

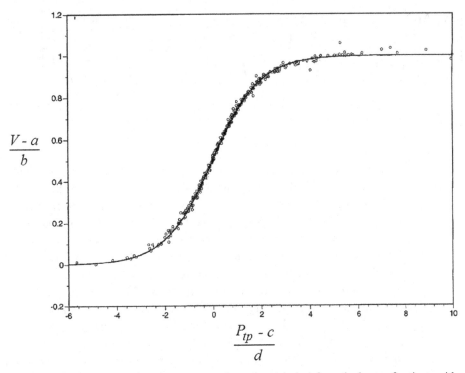

Figure 5.5 Normalized quasi-static pressure-volume data (circles) from the lungs of patients with acute lung injury and from dogs with a variety of pathologies, showing that the points collapse onto a single master curve described by Eq. 5.6 (line) (reproduced with permission from [89]).

5.2.2   Mechanisms of elastic nonlinearity

The finding that E_L depends on volume raises the question of what the genesis of this nonlinear behavior might be. We examined several mechanisms giving rise to elasticity itself in Section 4.3. Any one of these can give rise to nonlinear behavior. For example, Laplace's law for a fixed value of surface tension (Eq. 4.23) predicts that as the internal radii of curvature of airways and alveoli decrease, elastic recoil pressure, and hence E, should increase. This may contribute to the inverse relationship between E and transpulmonary pressure that is seen when lung volume descends below normal functional residual capacity (Fig. 5.5). However, the presence of pulmonary surfactant causes surface tension to vary with lung volume [90]. Furthermore, the complex behavior of surfactant molecules at the interface between air and liquid in the lungs also gives rise to hysteretic behavior, so the pressure at a given lung volume may be quite different during inflation compared to deflation. The amount of surfactant at the interface and its distribution throughout the alveoli and airways are also influenced significantly by the dynamic events of breathing [90]. The way that surface tension causes E to vary with lung volume is thus complicated and may give rise to significant nonlinear mechanical behavior.

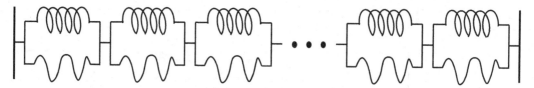

Figure 5.6 Spring-string model for predicting the quasi-static stress-strain behavior of lung tissue based on the notion of sequential recruitment of collagen fibers.

Another mechanism that can cause E to increase with decreasing lung volume is progressive closure of distal airways and alveoli. This may occur through the formation of *liquid bridges* across the lumen of small airways as a result of instabilities in the liquid layer that lines them [91]. Once formed, surface tension makes such bridges resistant to elimination, effectively *derecruiting* the downstream segment of the lung. Derecruitment may also occur through atelectasis [26] or accumulation of fluid in the alveoli [61]. Regardless of how it happens, derecruitment reduces the amount of lung tissue that can be ventilated, and is a function of both pressure and time [63, 92]. Consequently, derecruitment can cause very nonlinear elastic behavior in the lungs.

The lung tissue itself can also contribute to the volume dependence of E. In fact, nonlinear elastic behavior is typical of biological soft tissues, and represents what is known as *strain stiffening*. This is a familiar phenomenon. Try stretching any elastic material – a sheet of rubber, an elastic waistband – and you will invariably find that it goes only so far and then will stretch no further. Strain stiffening in any material ultimately reflects something about its structure, and in the case of the lung there is a generally accepted theory as to its main cause [93]. This theory concerns the arrangement of the two most important structural proteins in lung tissue, *elastin* and *collagen*. Both elastin and collagen are large macromolecules arranged into long fibers, but these fibers have very different mechanical properties. Elastin fibers are easily extensible and can be strained by 100% without rupturing [94]. Collagen fibers, on the other hand, are several orders of magnitude stiffer than elastin fibers, and can only be extended by a few percent of their unstressed lengths before they rupture [95–97].

The load-bearing component of lung tissue can be viewed in simplistic terms as a random meshwork of collagen and elastin fibers. At low strains (lung volumes) most of the collagen fibers are flaccid, exhibiting a wavy appearance under the microscope, so they are generally not under tension. The mechanical load on the tissue at low strains is thus borne mainly by the elastin fibers, which remain under some degree of tension. As strain increases, however, the slackness in the collagen fibers begins to be taken up, causing them to come under tension and eventually to take over the load-bearing role from the elastin. This happens gradually because not all the collagen fibers become taut at the same strain. The result is a gradual stiffening of the tissue.

This progressive recruitment of collagen fibers into the load-bearing role with increasing strain can be modeled using a collection of spring-string pairs arranged in series [98] as illustrated in Fig. 5.6. Suppose all the springs are identical with spring constant k, and that the strings have a range of different lengths, l_i. When a given tension, T, is applied across the entire model, the same T is experienced by each spring-string pair. For those pairs in which the string is sufficiently long, the associated spring is free to extend to the length T/k. For the remaining pairs the strings are taut and T is borne entirely by the string, which is infinitely stiff. If the spring-string pairs are sufficiently short and numerous then the various different string lengths can be assigned to a continuous distribution, $N(l)$. The total length L of the model thus comprises two components – those springs that are constrained to stay at their associated string lengths, and the remainder that are extended to length T/k. These two contributions, in addition to the unstressed length of the tissue, L_0, give the total length of the tissue as

$$L(T) = \int_0^{T/k} l N(l)\,dl + \int_{T/k}^{\infty} \frac{T}{k} N(l)\,dl + L_0. \tag{5.7}$$

Differentiating this expression with respect to T, using the standard calculus formula for differentiation of an integral, gives

$$\frac{dL(T)}{dT} = \frac{1}{k}\frac{T}{k} N\left(\frac{T}{k}\right) + \int_{T/k}^{\infty} \frac{1}{k} N(l)\,dl - \frac{1}{k}\frac{T}{k} N\left(\frac{T}{k}\right)$$

$$= \int_{T/k}^{\infty} \frac{1}{k} N(l)\,dl. \tag{5.8}$$

Differentiating again gives

$$\frac{d^2 L(T)}{dT^2} = -\frac{1}{k^2} N\left(\frac{T}{k}\right)$$

$$= -\frac{1}{k^2} N(l). \tag{5.9}$$

The string-length frequency distribution, $N(l)$, is thus proportional to the second derivative of length with respect to tension.

It turns out that the quasi-static force-length behavior of a strip of lung tissue can be accurately approximated by an increasing exponential relationship of the form

$$F(L) = \alpha(e^{\beta L} - 1) \tag{5.10}$$

where α and β are constants [98]. Applying Eq. 5.9 to this expression gives

$$N(l) = \frac{k^2}{\alpha^2 \beta} e^{-2\beta l}. \tag{5.11}$$

The model in Fig. 5.6 thus predicts that the distribution of string lengths is a decreasing exponential function of length.

A corresponding model can be developed by assuming that the strings are all of the same length, and that the values of the spring constant k are distributed instead. This gives rise to a similar second-order differential equation relating stiffness to the spring constant distribution, which when matched to experimental data predicts a peaked distribution for k with a long tail [98]. Interestingly, there is histological evidence that the widths of elastin fiber bundles in the lungs are indeed distributed in such a manner [97].

Collagen and elastin are not the only important structural components of lung tissue that potentially have bearing on its nonlinear behavior. Other key components include a hydrated gel, sometimes referred to as *ground substance*, composed of a variety of complex biomolecules [99]. The simple model presented above thus by no means constitutes a comprehensive description of lung tissue elasticity. Nevertheless, the model does serve to illustrate an important point: the mechanical behavior of lung tissue cannot be completely understood by considering the tissue components in isolation. Key aspects of these properties are *emergent*, meaning that they arise from interactions between tissue components. Accordingly, the ongoing quest for a deeper understanding of the complex nonlinear mechanical behavior of lung tissue is likely to rely increasingly on mechanisms of emergent behavior.

Another example of a mechanism for emergence that has recently been applied to the study of lung tissue mechanics is that of *percolation*, which describes the process whereby some agent attempts to traverse a labyrinth of obstacles and either makes it across the labyrinth or dies trying. The condition under which the agent just manages to make the traverse is known as the *percolation threshold*. Percolation has been used to model an enormous range of natural processes, and in fact appears in a two-dimensional version of the series spring-string model shown in Fig. 5.6 [100]. In the two-dimensional model at low strain, the network elements that are stiffened by the stretching of collagen are few and far between, and are therefore isolated from each other by a sea of flaccid collagen fibers. However, as strain increases there comes a point when, all of a sudden, a contiguous path of stiffened elements percolates across the network from one side to the other. This marks the percolation threshold at which point the bulk stiffness of the entire network suddenly starts to increase at a much greater rate [100].

Percolation can also be used to understand how the progression of pathology in lung parenchymal disease might become disconnected from the progression of symptoms [101]. Again, consider the parenchyma of the lung as being like a network of nodes connected by springs, and suppose that the springs initially all have identical stiffness. The bulk modulus of the network can be determined by stretching its bordering nodes by a small amount and then determining the mean increase in tension across the network (the technique of simulated annealing can be used to determine the elastic equilibrium configuration of the internal nodes for any given boundary configuration [100, 101]). Next, randomly increase the stiffness of individual springs to represent progressive fibrosis of the lung. The stiffness of the network increases monotonically (Fig. 5.7) but there is a sharp increase in its slope at the percolation threshold, which is reached when a contiguous chain of stiffened springs first spans the network. This model thus predicts that patients with pulmonary fibrosis may continue to progress in their disease for a while with modest progression in symptoms until they reach the percolation threshold. At this point, a sudden dramatic decline can be expected, which seems to match at least some clinical reports [102].

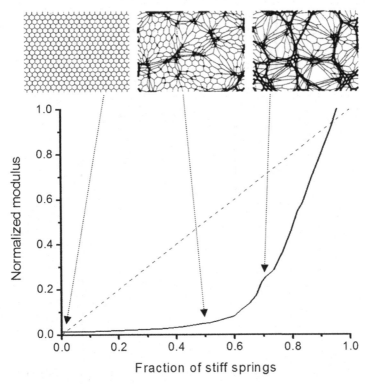

Figure 5.7 Simulation of the progression of pulmonary fibrosis using a two-dimensional network of Hookean springs. The solid curve shows the bulk modulus of the elastic network versus the fraction of springs randomly stiffened by a factor of 100 (normalized to the modulus of the network when all springs are stiffened). The network configurations obtained when 0%, 50%, and 67% of the springs have been stiffened are linked to their respective positions on the modulus plot by the dotted arrows, and show that the percolation threshold occurs between 50% and 67%. If all of the spring constants are uniformly stiffened in a gradual manner from the baseline value of 1 to 100, the modulus follows the dashed diagonal line (reproduced with permission from [101]).

Models such as those illustrated in Figs. 5.6 and 5.7 are, of course, highly oversimplified representations of reality. Nevertheless, they likely capture some key aspects of the way that the macroscopic behavior of the lung arises from its constituents. Importantly, these models highlight the fact that much about the macroscopic mechanical properties of lung tissue is emergent.

5.3   Choosing between competing models

The models considered above all have a single mechanical degree of freedom, so they are based on the assumption that the lungs are ventilated homogeneously. In other words, each region of the lung receives the same fraction of the tidal volume, in proportion to

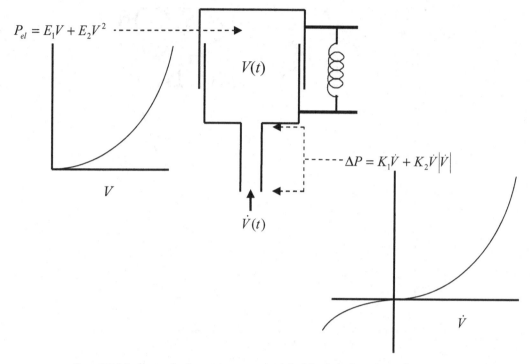

$$P_{el} = E_1 V + E_2 V^2$$

$V(t)$

$$\Delta P = K_1 \dot{V} + K_2 \dot{V} |\dot{V}|$$

V

$\dot{V}(t)$

$\dot{V}$

Figure 5.8 Nonlinear single-compartment model of the lung featuring a flow-dependent resistance and a volume-dependent elastance.

its initial volume, as every other region. These models thus extend the linear model of Fig. 3.1 merely by invoking nonlinear constitutive properties for either the airway or the lung tissue. Of course, both nonlinearities could be present at the same time, as described by the union of Eqs. 5.2 and 5.4. That is,

$$P_{tp} = K_1 \dot{V} + K_2 \dot{V} |\dot{V}| + E_1 V + E_2 V^2 + P_0, \tag{5.12}$$

as illustrated in Fig. 5.8.

Even the model in Fig. 5.8, however, does not account for all the deterministic variation in either of the two patient data sets shown in Fig. 5.1. One way to try to deal with this situation would be to make R_L and E_L even more complicated. For example, R_L could be described by a higher-order polynomial in $\dot{V}$ and E_L by another polynomial in V [37]. A further possibility would be to let R_L vary inversely with V to account for the increase in airway caliber that occurs as the lungs are inflated. In principle, this process of adding terms to the equation can be pursued without limit, adding progressively higher-order polynomial terms in V, $\dot{V}$, and their products until all obvious deterministic variation in the data is accounted for and all that is left is random noise. There is no guarantee, of course, that all these additional parameters will be interpretable in a way that leads to physiological insight. In fact, the relentless pursuit of an improvement in fit can result in a model that has no clear interpretation at all. However, even if a good fit is all that matters, one still has to know when the fit is good enough to stop the process of adding more parameters. To address this question it is necessary to go back to the assumptions on which model fitting is based.

The conventional view of the model-fitting process is that the dependent variable (P_{tp} in this case) consists of two independent parts, a *deterministic* part and a *stochastic* (random) part. The deterministic part is presumed to reflect the processes we are interested in, which are presumably attributable to the laws of classical physics. The stochastic part, on the other hand, is assumed to be noise, that ubiquitous nuisance reflective of our inability to measure things with perfect accuracy. If we fit a model to P_{tp} which describes all of its deterministic parts, then the residuals (the parts of P_{tp} that the model cannot account for) ought to have the character of random noise. Conversely, if the model is not complicated enough to describe all the deterministic parts of the data, then the residuals will not be randomly scattered but instead will exhibit systematic fluctuations either side of zero.

The pattern of residuals thus holds the key to deciding whether one model or another is appropriate for a given set of respiratory data. Visual inspection of these patterns is sometimes all that is needed to decide whether a particular model should be retained or discarded in favor of a competitor. However, in those cases where the choice is not obvious, an objective measure of relative model performance is needed. Indeed, we face precisely this problem with the nonlinear models of the lung developed above.

The linear single-compartment model fit the data from Patient B in Fig. 5.1 reasonably well, and when this model was extended to include either a flow-dependent resistance or a volume-dependent elastance, it gave an improved fit as evidenced by the smaller residuals in Figs. 5.2 and 5.3, respectively. Simply looking for a decrease in the *MSR*, however, is not enough to justify adoption of a more complicated model. A model with more parameters will always fit the data better. After all, if a model has n free parameters, it should fit n data points perfectly regardless of how much noise the points contain. The key issue, therefore, is whether the more complicated model fits the data better by a big enough margin to justify its adoption over the simpler model.

5.3.1   The *F*-ratio test

A mathematical model is a *hypothesis*, a quantitative statement about how a system is presumed to work. Accordingly, choosing between competing models can be viewed as a matter of *hypothesis testing*. This amounts to deciding whether one model fits a given data set better than another model in a statistically significant sense. To make such a decision, we need to look for a change in some quantity that conforms to a known *probability distribution function*. The regression theory outlined in Section 3.2.2 points to the obvious candidate for this quantity, namely the *MSR*. The *MSR* is the variance of the residuals between a data set and its model fit. If the model accounts for all the deterministic variation in the data, and if the noise in the data is random and conforms to a *normal (Gaussian) distribution*, then the *MSR* is an estimate of the variance of that distribution. Accordingly, if two different models do an equally good job of fitting a given set of data, the values of the *MSR* from each fit constitute two estimates of the variance of the particular normal distribution from which the noise is derived.

Now, it so happens that if you take two random samples from a Gaussian distribution, one sample of size n and the other of size m, then the ratio of the variances of the two samples follows an F *distribution* [103]. That is,

$$\frac{\text{var}_1}{\text{var}_2} \in F_{n-1,m-1} \tag{5.13}$$

where var_1 and var_2 are the variances of the two populations, $n-1$ and $m-1$ are their respective numbers of degrees of freedom, and the ratio in Eq. 5.13 is always calculated with the larger variance in the numerator. By analogy, we can say that the ratio of MSR values from two different model fits should also follow an F distribution provided the models each account for all the deterministic variation in the data. The equivalence of the two models can then be tested under the *null hypothesis* that they are indeed equivalent. By convention, the null hypothesis is accepted if the ratio of MSR_1 to MSR_2 lies within the central 95% of the area encompassed by the F distribution. If the ratio is larger than the 95% value, that is

$$\frac{MSR_1}{MSR_2} > F_{n-1,m-1,0.05}, \tag{5.14}$$

then the null hypothesis is rejected and the *alternative hypothesis* is accepted. The alternative hypothesis is that the two models are not equivalent in their ability to describe the data, so the model providing the smaller value of MSR is taken to provide a statistically significantly better fit.

The F-ratio test can be used to compare the linear and nonlinear models of lung mechanics in terms of their abilities to fit the data from Patient B in Fig. 5.1. If we evaluate Eq. 5.14 using the MSR value from the linear model as MSR_1 and that from the flow-dependent resistance model (Eq. 5.2) as MSR_2 we obtain a value of $2.759/1.362 = 2.026$. The number of degrees of freedom associated with each fit is equal to the number of data points (3.7 s at 120 Hz = 444 points) minus the number of model parameters (3 and 4, respectively). Tables of the F distribution, which can be found in almost any text on statistics, typically provide critical values up to 120 degrees of freedom and then jump straight to infinity. The value of $F_{120,120,0.05}$ is 1.35, while $F_{\infty,\infty,0.05}$ is 1.22. The value pertinent to the lung model fits in Figs. 5.1 and 5.2 lies between these two values, but in any case is considerably less than the value of 2.026 calculated above. This means that, according to the F-ratio test, the improvement in fit in going from Eq. 3.5 to Eq. 5.2 is statistically significant. On the basis of the F-ratio test, therefore, for the data from Patient B we would discard the linear single-compartment model in favor of its extension that features a flow-dependent R. The same argument can be made with respect to the model with a volume-dependent E (Eq. 5.4 and Fig. 5.3).

There is an important caveat here, however. The F-ratio test is based on the assumption that the residuals between data and model prediction are random, uncorrelated, normally distributed noise. Figures 5.1 to 5.3 clearly show this not to be the case. Indeed, the residuals from all fits are so systematic that one would be hard pressed to argue they have much of a stochastic component at all. As pointed out in Section 3.3.2, in a case like this it makes little sense to speak of statistical significance when discussing fitting performance. One therefore has to take the results of the F-ratio test in circumstances such as these with something of a "grain of salt." Nevertheless, the F-ratio test can still provide some sense of whether or not one can take an improvement in model fit seriously.

5.3.2   The Akaike criterion

An alternative method for choosing between competing models is based on the *Akaike information criterion* (*AIC*) [104]. This approach seeks to identify the model that has the maximum statistical likelihood of describing the data, where likelihood is calculated on the basis of maximum entropy. The *AIC* was originally developed for situations in which n is large. A widely used corrected form, denoted *AICc*, can accommodate small values of n, and is defined as [105]

$$AICc = 2m + n\left[\ln\left(2\pi\, MSR\right) + 1\right] + \frac{2m(m+1)}{n-m-1}. \tag{5.15}$$

Increasing m always reduces MSR, which decreases the middle term on the right-hand side in Eq. 5.15. On the other hand, the other two terms in the equation both increase when m increases. *AICc* thus balances goodness-of-fit against an increase in the number of model parameters. The preferred model is the one with the lowest value of *AICc*.

An advantage of the *AICc* over the F-ratio test is that it allows multiple models to be compared at the same time [106]. Thus, for example, the values of *AICc* for the linear model (Eq. 3.5), the flow-dependent R_L model (Eq. 5.2), and the volume-dependent E_L model (Eq. 5.4) obtained with the data from Patient B are 1717, 1405, and 1332, respectively. This suggests that the volume-dependent E_L model is the best for describing this particular data set. It must be remembered, however, that "best" here means only with respect to the three models being compared. There might be an even better model we have not found yet that produces an even lower value of *AICc*.

Problems

5.1   Assuming Poiseuille flow in the airways, how would you expect airway resistance to depend on lung volume (assuming an isotropically expanding lung)? Write an equation of motion for a lung model having such a resistance. How would you estimate the parameters of this model from measurements of pressure, flow, and volume at the airway opening?

5.2   The value of the exponent K in Eq. 5.5 has been used as a diagnostic index of diseases of the lung parenchyma that affect its elastic properties. How might K change from normal in either emphysema or pulmonary fibrosis?

5.3   Adapt Eq. 5.4 so that the elastic properties of the lungs are described by Eq. 5.5. Do the same using Eq. 5.6. Which of the model parameters can be evaluated by multiple linear regression? Which parameters require a nonlinear optimization method?

5.4   Suppose the spring–string pairs in Fig. 5.6 are arranged in parallel instead of in series. Derive the equation relating tension to length (i.e. the counterpart to Eq. 5.9).

5.5 How might the concept of percolation be used to direct treatment of parenchymal diseases such as pulmonary fibrosis or emphysema?

5.6 Generate points along a straight line and add random numbers to the points to generate a noisy set of correlated data. Fit both a straight line and a second-order polynomial to these data. Calculate the F-ratio statistic (Eq. 5.14) and test for significance. Calculate the corrected Akaike criterion (Eq. 5.15) for the two fits and determine which is best according to this.

6   Flow limitation

So far, we have been systematically building up a quantitative mechanical view of the lung in terms of inverse models. Before continuing on in this upward climb toward increased model complexity, we will take a slight digression to consider in this chapter the phenomenon of expiratory flow limitation [107, 108]. As pointed out in Section 1.3, expiratory flow does not increase indefinitely as the expiratory muscles increase their forces of contraction. Instead, expiratory flow approaches a maximum value that can not be exceeded regardless of how much extra effort is exerted. It is convenient to think of flow limitation as yet another kind of nonlinearity that can afflict the single-compartment model, and so its discussion here makes a natural follow-on from the previous chapter. However, flow limitation has special status in terms of clinical application, so no treatise on lung mechanics would be complete without giving it due consideration. The following development is based on [109] and [110].

6.1   FEV$_1$ and FVC

It seems natural to think that the harder you try to force air out of your lungs, the faster it will come out. While this is certainly the case at low rates of flow, it is not generally true. As expiratory effort is increased, there comes a point at which further increases in effort are not rewarded with increases in flow. This *maximum expiratory flow* is strongly influenced by the tethering forces of the parenchyma that pull outward on the airway wall. These forces stabilize the airway against those factors that act to limit flow, which causes maximum expiratory flow to increase markedly with increases in lung volume. As a result, maximum flow is beyond the reach of the respiratory muscles at lung volumes near total lung capacity even in a healthy subject. However, over the lower 80% or so of the vital capacity range, a normal individual can generate sufficient expiratory muscle force to reach maximal flow [107].

The shape and position of a plot of flow ($\dot{V}$) versus volume (V) during a maximal expiratory effort is diagnostic of lung disease. The most commonly used parameter derived from the phenomenon of flow limitation is the volume of air expired in the first second of a forced expiration following a maximal inspiration, known as FEV$_1$. Maximal flow decreases with decreasing airway radius, which explains why individuals with narrowed airways (for example, subjects with asthma or chronic obstructive pulmonary disease) have a reduced FEV$_1$ compared to normal individuals, as discussed in

Section 1.3. Maximal flow also decreases as the airway wall becomes less stiff. FEV_1 is thus reduced in subjects with abnormally compliant airways, such as in emphysema [111].

FEV_1 is a sensitive parameter diagnostic of lung disease, but its specificity is limited by the fact that a lower-than-expected value can indicate either that the airways are not as patent as they should be (as in an obstructive pulmonary disease such as asthma) or that there was a problem in expanding the lungs properly in the preceding maximal inspiration (as in a restrictive pulmonary disease such as pulmonary fibrosis). FEV_1 is even more powerful diagnostically when coupled with a measure of the total volume of gas expired during an entire forced expiration, known as forced vital capacity (FVC). Typically, both FEV_1 and the ratio FEV_1/FVC are lower than normal in obstructive lung disease, although FVC may also be reduced due to complete collapse of some airways before expiration is complete. This traps gas within the lungs. In restrictive disease, FVC is lower than normal but FEV_1/FVC tends to be preserved [107, 108, 112].

The forced expiratory maneuver has several great advantages from a clinical standpoint. It is measurable with simple and relatively inexpensive equipment. It is also completely noninvasive and requires only that the subject be able to inhale maximally and then exhale forcibly. The maneuver is thus applicable to all patients except young children (typically less than five years of age) and patients with certain neuromuscular pathologies. For the same reason, however, forced expiratory maneuvers have limited utility as a research tool in laboratory animals.

The major problem with using maximum expiratory flow diagnostically is that linking changes in FEV_1 and FVC to abnormalities in lung structure is not straightforward. Some computational modeling studies have produced realistic simulations of maximal expiratory flow [113, 114], but the models are complicated and have not yet found wide acceptance. In any case, to link structure to function it is necessary to understand the physics underlying the phenomenon of expiratory flow limitation. In this chapter, we will review the basics of this physics by deriving the important concepts from first principles.

6.2   Viscous mechanisms

At the simplest level, flow limitation can be understood in terms of the viscous effects of gas flowing along a collapsible airway [109]. Suppose the alveolar regions of the lung behave as a single compartment exiting through a single conduit representing the conducting airways. The expiratory muscles act to force air from this structure by applying a positive pressure (pleural pressure, P_{pl}) to both the alveolar compartment and the outside of the airway (Fig. 6.1). The alveolar compartment is compliant, so an increase in P_{pl} causes an increase in alveolar pressure (P_A) relative to the pressure at the airway opening (P_{ao}). The result is a pressure gradient that drives flow out along the airway. If the airway is rigid, the rate at which air flows along the airway ($\dot{V}$) will increase, in principle without limit, as P_A increases. In such a situation, there is no finite maximal flow.

The situation is different, however, if the conduit is compliant, because now it will tend to narrow in the presence of an inward-acting transmural pressure across its walls. The

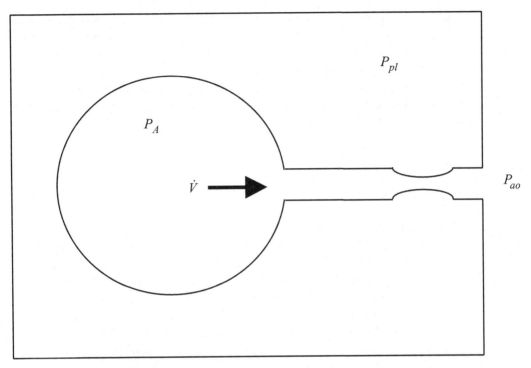

Figure 6.1 Stylized model of the respiratory system featuring an elastic alveolar unit, internal pressure P_A, emptying through an airway to an outside pressure of P_{ao}. Both the alveolar unit and airway are encased in a chest wall compartment that applies a pressure P_{pl} to both (adapted with permission from [110]).

pressure inside the conduit decreases from P_A at its distal end to P_{ao} at its proximal end because of the viscous pressure drop present in the flowing stream of air. Consequently, the transmural pressure increases along the airway from $P_{pl} - P_A$ to $P_{pl} - P_{ao}$ (Fig. 6.1). An increase in P_{pl} produces two opposing effects: it increases the upstream driving pressure, which tends to increase $\dot{V}$, and it narrows the conduit, which increases its resistance and decreases $\dot{V}$. If the latter effect ever gets to the point of matching or exceeding the former, $\dot{V}$ will not be able to increase despite increases in P_{pl}. Flow thus becomes limited when any further increases in the driving pressure along the airway are offset by increased luminal narrowing. The precise point along the airway where this limiting process takes place in an actual lung (the so-called *choke point*) varies with the mechanical properties of the airways and parenchyma, and moves from the large extrapulmonary airways at high lung volume to the lung periphery as volume decreases [115].

6.3   Bernoulli effect

The explanation just given for flow limitation is based on the flow resistance of the airway, which arises as a consequence both of the viscosity of air (when flow is laminar) and air density (when flow becomes turbulent). This is not the only way of getting flow limitation, however. An alternative explanation is based on the *Bernoulli effect*, which depends only on the density (ρ) of air. As discussed in Section 2.2.2, when air flows at a rate $\dot{V}$ along a conduit of cross-sectional area A, the pressure at any point in the conduit measured perpendicular to the direction of flow (the so-called lateral pressure) is less than the pressure driving flow at that point by an amount (P_b) proportional to the axial flow velocity (v) thus:

$$P_b = \frac{\beta \rho v^2}{2}$$
$$= \frac{\beta \rho \dot{V}^2}{2A^2} \tag{6.1}$$

where β is a constant that depends on the flow velocity profile [5, 84]. We will assume from now on that β equals 1, which corresponds to uniform flow (i.e. velocity is independent of radial position in the conduit). The Bernoulli effect is a manifestation of the conservation of energy; because fluid mass (i.e. volume) is conserved, narrowing the conduit for a given $\dot{V}$ will cause v to increase [72]. This is achieved, however, without any addition of external energy, so must be offset by a decrease in the potential energy of the fluid, which is reflected in a decrease in its lateral pressure.

As air flows along a compliant airway, the decreased lateral pressure due to the Bernoulli effect results in an increased transmural pressure. This effectively sucks the airway walls inward, which narrows the airway lumen (i.e. reduces A in Eq. 6.1). For a given $\dot{V}$, this narrowing increases v, which further decreases lateral pressure, which in turn further decreases A. By this mechanism it is possible, depending on the elastic characteristics of the airway wall, to have complete collapse of the airway. As soon as this happens, of course, flow ceases altogether, so the Bernoulli effect disappears and the airway opens up again, only to have the process repeat itself. This may lead to audible airway flutter.

To see how the Bernoulli effect can lead to flow limitation, suppose that the flexible conduit carrying the flow has a linear relationship between its transmural pressure (P_{tm}) and A. The slope of the relationship is dP_{tm}/dA, so

$$P_{tm} = \frac{dP_{tm}}{dA}(A_0 - A) \tag{6.2}$$

where A_0 is the tube area at zero P_{tm}. If we neglect viscous effects (i.e. let airway resistance be zero) then P_{tm} is equal simply to P_b, as given by Eq. 6.1. We can thus write

$$\frac{\rho \dot{V}^2}{2A^2} = \frac{dP_{tm}}{dA}(A_0 - A) \tag{6.3}$$

or

$$\dot{V} = \sqrt{\frac{2A^2}{\rho}\frac{dP_{tm}}{dA}(A_0 - A)}. \tag{6.4}$$

This equation defines the functional relationship between $\dot{V}$ and A. Inspection of Eq. 6.4 reveals that $\dot{V}$ equals zero when A is equal to either zero or A_0. Furthermore, when A starts to become less than A_0 (i.e. when the airway starts to narrow) the above expression starts to become positive. This means that the magnitude of $\dot{V}$ must reach a maximum for some value of A between zero and A_0. The slope of this relationship is zero at the maximum, so we can find the exact value of A at which it occurs by differentiating $\dot{V}$ with respect to A and setting the result equal to zero. The derivative of Eq. 6.4 is

$$\frac{d\dot{V}}{dA} = \frac{(2A_0 - 3A)\dfrac{A\,dP_{tm}}{\rho\;dA}}{\sqrt{\dfrac{2A^2\,dP_{tm}}{\rho\;dA}(A_0 - A)}}$$
$$= 0. \qquad (6.5)$$

Making this expression equal zero is simply a matter of choosing a value for A that makes the numerator zero. There are two possible choices; either A is zero, for which $\dot{V}$ is itself zero, or the quantity $(2A_0 - 3A)$ is zero in which case

$$A = \frac{2}{3}A_0. \qquad (6.6)$$

Using this expression to replace A_0 in Eq. 6.4 provides the value for the maximal $\dot{V}$ as

$$\dot{V}_{max} = A\sqrt{\frac{A\,dP_{tm}}{\rho\;dA}}. \qquad (6.7)$$

This equation, which can be derived in a more general setting [116], shows how the Bernoulli effect imposes a certain functional dependence of $\dot{V}_{max}$ on airway area, the density of air, and the elastic behavior of the airway wall. It is clear from this equation, for example, that a decrease in airway stiffness (i.e. dP_{tm}/dA) will reduce $\dot{V}_{max}$ as in emphysema, while an increase in A will increase $\dot{V}_{max}$, as occurs with bronchodilatation. It is also clear from Eq. 6.7 that $\dot{V}_{max}$ should be greater when a subject breathes a less dense gas, as is indeed the case with a mixture of helium and oxygen [117].

6.4 Wave speed

Another concept that is considered central to the determination of maximal expiratory flow is that of *wave speed* [116]. The wave speed theory of flow limitation was originally introduced to the respiratory community as an adaptation of the concept applied to the flow of urine along the urethra [116]. To understand wave speed theory, it is instructive to see how the equations governing wave propagation can be derived from first principles [118]. First, consider the flow of air through a rigid conduit. (Actually, the walls of an airway are not perfectly rigid, but what is important here is that they are rigid compared to the compressibility of the gas inside the conduit.) If an aliquot of air is introduced at one end of the conduit, the pressure at that end increases accordingly, which sends a pressure pulse to the other end at the speed of sound. This establishes the pressure gradient necessary to drive the aliquot along the conduit. For a structure the size of a human lung, this is effectively instantaneous compared to the time scales of normal respiratory events. Furthermore, there is, in principle, no limit to how big the aliquot of air can be, and thus no limit to the flow that is produced.

Now consider a conduit with walls that are compliant compared to the air inside it, so we may consider the air to be incompressible. When an aliquot of air is introduced into such a conduit it is initially accommodated by a lateral expansion of the conduit's upstream walls. This expansion then propagates toward the other end of the conduit at a rate determined not by the speed of sound, but by the much lower speed of movement of the elastic deformation of the wall. To accommodate a steady stream of air in this way, the conduit walls must move in a continuous wave, meaning that each point in the wall must oscillate radially either side of its relaxed position. Because the air is moved along by the wave motion of the conduit walls, the velocity of propagation of the air equals the speed of these elastic waves. The actual flow transmitted is given by the product of this velocity and the amplitude of the oscillations. Obviously, once these oscillations reach an amplitude equal to the conduit radius, the opposing walls will bump into each other at the peak of their inward excursions, thereby limiting the flow that can be carried in this way.

To see how this works mathematically, suppose we have a perfectly cylindrical compliant airway. From the circular symmetry of the situation, we can reduce the problem to a single dimension and consider oscillations in an elastic string representing a thin strip of width α cut axially along the airway (Fig. 6.2A). Let the relaxed (equilibrium) position of this string be at a distance r_0 from the center of the airway, and the axial distance along the string be x (Fig. 6.2B). Suppose, now, that a parcel of air is positioned so as to displace a portion of the string laterally from its equilibrium position, so that the radius r becomes a function of x. As the string is actually part of an elastic tube, this generates a tension of magnitude αT in the displaced portion of the string, where T is the tension per unit circumferential distance around the tube wall. We will assume that the tension at each point along the string is proportional to the magnitude of the lateral displacement (i.e. $r - r_0$) at that point. That is,

$$\alpha T = \alpha E(r - r_0) \tag{6.8}$$

where E is the specific elastance of the airway wall.

We now focus our attention on the infinitesimal region between axial positions x and $x + dx$ (Fig. 6.2C). The tension in the string at position x has a vertical component equal to $\alpha T \sin\theta$, which acts to pull down on the segment of string between x and $x + dx$. Similarly, an upward force is applied by the vertical component of tension at position $x + dx$. The net force (F) acting upward between x and $x + dx$ is thus the difference between the vertical components of tension at $x + dx$ and x. If the slope of the string at both positions is small (i.e. the height of the parcel of air is small compared to its length) then $\sin\theta$ is approximately equal to the derivative of r with respect to x, so we can write

$$F = \alpha E(r - r_0)\left[\left(\frac{\partial r}{\partial x}\right)_{x+dx} - \left(\frac{\partial r}{\partial x}\right)_x\right] \tag{6.9}$$

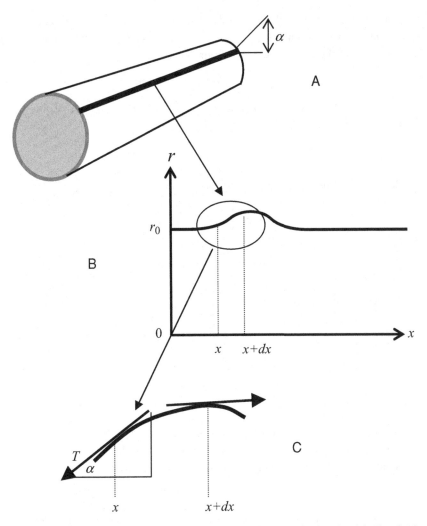

Figure 6.2 (A) A thin string of width α is cut axially from an elastic tube. (B) The elastic string is displaced radially from its equilibrium position of r_0. (C) The string has tension at any point proportional to its displacement from r_0 at that point. The tension in the string at positions x and $x + dx$ has vertical components whose difference gives the net force acting on the intervening section (reproduced with permission from [110]).

where we have used partial derivatives to denote the slope of the string because the position of each point along the string (i.e. the function r) varies with both x and t. Note that the above expression is only an exact equality because dx is infinitesimal. If dx were of finite extent it would be only approximately true. Note, also, that we are only considering radial displacements of the string. When displacements are large, it is possible to have a point on the string move axially as well, but we will ignore these effects here.

F acts vertically on the mass of the string between x and $x + dx$ and on the small column of air between the string and r_0. We will assume the string mass to be negligible compared to the mass of the air column, which is

$$M = \alpha(r - r_0)\rho dx \tag{6.10}$$

where ρ is the density of air. The radial acceleration, $\partial^2 r/\partial t^2$, of this mass of air is provided by F according to Newton's law of motion

$$F = M\frac{\partial^2 r}{\partial t^2}. \tag{6.11}$$

Substituting Eqs. 6.9 and 6.10 into Eq. 6.11 and rearranging gives

$$\frac{\left[\left(\frac{\partial r}{\partial x}\right)_{x+dx} - \left(\frac{\partial r}{\partial x}\right)_x\right]}{dx} = \frac{\partial^2 r}{\partial x^2}$$

$$= \frac{\rho}{E}\frac{\partial^2 r}{\partial t^2}. \tag{6.12}$$

This equation can be expressed as

$$\frac{\partial^2 r}{\partial t^2} = v^2 \frac{\partial^2 r}{\partial x^2} \tag{6.13}$$

where

$$v = \sqrt{\frac{E}{\rho}}. \tag{6.14}$$

Equation 6.13 involves derivatives of r with respect to both x and t, and is therefore a partial differential equation. In fact, it is the one-dimensional version of the wave equation. Fortunately, this equation can be solved analytically. The solution is

$$r(x, t) = r(x \pm vt) \tag{6.15}$$

meaning that r can be any function you like provided that the variables x and t always appear together in the combination $x - vt$ or $x + vt$. Equation 6.15 thus describes a traveling wave, a function that moves either to the right (if the argument is $x - vt$) or the left (if the argument is $x + vt$) without changing shape. Furthermore, the speed with which this wave moves is v. To see this, notice that if a wave moving to the right starts out at $t = 0$ being described by the function $r(x)$, at a time t seconds later it is described by $r(x - vt)$. This is still the same function r (i.e. it has not changed shape), but it has been displaced to the right by the distance vt, so v must be its velocity.

Finally, we want to express v in terms of the variables related to flow through a conduit, namely the cross-sectional area (A) of the conduit and P_{tm}. First, we note that

$$E = \frac{dT}{dr}. \tag{6.16}$$

Since $A = \pi r^2$, we have

$$dA = 2\pi r dr. \tag{6.17}$$

Next, we recall that the pressure in a sphere due to surface tension in its curved walls is given by Laplace's law (Eq. 4.22) as $P_{tm} = 2T/r$. The equivalent relationship for a cylinder is $P_{tm} = T/r$. Assuming that the changes in r involved are small, this gives

$$dP_{tm} = \frac{dT}{r}. \qquad (6.18)$$

Substituting Eqs. 6.17 and 6.18 into Eq. 6.16, and then putting the result into Eq. 6.14, gives the wave speed as

$$v = \sqrt{\frac{\pi r^2}{2\rho} \frac{dP_{tm}}{dA}}$$

$$= \sqrt{\frac{A}{2\rho} \frac{dP_{tm}}{dA}}. \qquad (6.19)$$

Once the amplitude of elastic waves in the string considered above reaches r_0, the inward movement of opposing walls prevents further excursions. This is equivalent to having the circular cross-section of the tube going to zero at the nadir of each oscillation. The mean cross-sectional area of the tube as its radius oscillates sinusoidally between 0 and $2r$ is about $1.5A$. This gives a maximal flow of

$$\dot{V}_{max} = 1.5A\sqrt{\frac{A}{2\rho} \frac{dP_{tm}}{dA}}$$

$$\approx A\sqrt{\frac{A}{\rho} \frac{dP_{tm}}{dA}}.$$

Note, however, that such large excursions of the airway wall are at odds with the requirement invoked in the above derivation that changes in r be small, so Eq. 6.20 can only be an approximation of what actually happens. In any case, Eq. 6.20 is the formula for the $\dot{V}_{max}$ predicted by wave speed theory [107, 109].

The limiting effects of wave speed can also be understood from the perspective of what happens when the downstream pressure in an airway is lowered. In order for the lowering of downstream pressure to have an effect on $\dot{V}$, the mechanical perturbations produced by the lowering of the pressure must be transmitted to the upstream end of the airway. In other words, an increase in $\dot{V}$ can only occur if the delivery of air to the upstream end is increased, and this only takes place if the upstream end knows that pressure has been lowered at the downstream end. However, this transmission of information takes place at wave speed, so if the gas is already flowing downstream at this rate, the perturbations never have a chance to reach the upstream end of the airway [109]. Consequently, the upstream pressure is never affected by anything that happens downstream. This is reminiscent of water flowing over a waterfall; lowering the downstream level of the waterfall does nothing to change the flow over the waterfall itself [119].

An interesting thing about the $\dot{V}_{max}$ specified by Eq. 6.20 is that it is exactly the same as the $\dot{V}_{max}$ determined by Eq. 6.7, implying that wave speed and the Bernoulli effect are expressions of the same phenomenon. Indeed, both are descriptions of the effects of convective acceleration of gas along the airway based on Newton's law and the conservation of mass and energy. In a real lung, both the Bernoulli/wave speed mechanism and the viscosity-based choke point mechanism play roles in determining limiting flow. Indeed, it is thought that the viscous mechanism dominates at low lung volumes, while the density-dependent mechanism is more important at high lung volumes [107].

Finally, it is important to remember that the various flow-limiting mechanisms described above were developed on the basis of a number of rather substantial assumptions. In the case of wave speed, for example, it was assumed that the amplitude of the elastic wave is small compared to its wavelength, the radial excursions of the airway wall are small, the air in the airway is incompressible and inviscid, and the airway wall is linearly elastic and massless. These assumptions are, of course, not realized in a real lung. Also, the airways of the lung are not a single conduit but rather comprise a heterogeneous branching tree structure. Some of these assumptions may not even be close to being satisfied, such as the small amplitude requirement. More accurate predictions of $\dot{V}_{max}$ could be obtained by relaxing some or all of the assumptions invoked in deriving the above equations, but the mathematics involved rapidly becomes rather complicated [120]. Even so, the wave speed formula has been applied with some success to anatomically accurate computational models of the airway tree that appear to give reasonable predictions of the phenomenon of flow limitation [113, 114]. In any case, the theories outlined above serve to delineate those factors, such as airway caliber, airway wall stiffness, and gas density, that play the major roles in determining maximal expiratory flow. As such, they are invaluable in providing physical insight into a process whose importance for clinical respiratory medicine cannot be overestimated.

Problems

6.1 How would the magnitude of maximal expiratory flow be affected by the velocity profile of flow in the airways?

6.2 If we breathed water rather than air, would our maximal expiratory flows be greater or less? Explain.

6.3 The wave speed formula of flow limitation (Eq. 6.20) is based on the assumption of small wave amplitudes. How would you expect wave speed to be affected by relaxing this assumption and allowing wave amplitude to be significant compared to wavelength?

6.4 How would the expiratory flow-volume curve be affected by (a) emphysema and (b) pulmonary fibrosis?

6.5 In this chapter we discussed the phenomenon of expiratory flow limitation. However, flow limitation during inspiration can also be a problem in some situations. What might these situations be?

6.6 Expiratory flow limitation only happens in a healthy lung during a forced expiration. In some diseases, however, it can occur during normal tidal breathing. The prime example of this is chronic obstructive pulmonary disease (COPD), during which the airways of the lung may become so floppy that they collapse during normal breathing. To compensate for this, patients often tend to breathe at higher than normal lung volumes. Why would this tend to offset the tendency for the airways to dynamically collapse during expiration?

7   Linear two-compartment models

We now come back to the general question of how the single-compartment linear model can be extended to provide a more realistic representation of the lung. Even this simple model has proven extremely useful for describing the behavior of the lung (Chapters 3 and 4). It does even better when extended to include nonlinearities related to flow and volume (Chapters 5 and 6). Nevertheless, a model with only one alveolar compartment just seems too simplistic. It is obvious from considerations of anatomy that a real lung can never be perfectly homogeneous, even when healthy. Structural asymmetries in the airway tree alone make it easier for inspired air to reach some alveolar regions than others. It is also hard to imagine that any disease process affecting mechanical function would strike the lung in a geographically uniform manner. It is therefore natural to think that a realistic model of lung mechanics should account for regional differences in mechanical properties. Indeed, one does not have to look far to find examples of significant departure from single-compartment behavior in the lung.

7.1   Passive expiration

One source of experimental evidence for the presence of more than one compartment in the lung comes from *passive expiration*. Here, the lungs are inflated to some volume V_0 above functional residual capacity, and then suddenly allowed to exhale solely under the influence of the elastic recoil forces of the respiratory tissues. If expiration begins at time $t = 0$, then according to Eq. 3.3 the equation describing the subsequent evolution of the system is

$$\dot{V}(t) = -\frac{E}{R}V(t) \tag{7.1}$$

subject to the initial condition $\dot{V}(t) = V_0$. The solution to this equation is a single decaying exponential function of time. That is,

$$V(t) = V_0 e^{-t/\tau} \tag{7.2}$$

where τ is the time-constant of the model, given by

$$\tau = \frac{R}{E} \tag{7.3}$$

as can be seen by substituting Eq. 7.2 into Eq. 7.1.

A test of the validity of the single-compartment linear model would therefore be to see if Eq. 7.2 accounts for measurements of $V(t)$ recorded during a passive expiration, a relatively simple experiment. The process of fitting Eq. 7.2 to data is not quite as simple as finding E and R for the single-compartment model because τ is nonlinearly related to the dependent variable $V(t)$. This model fitting problem is thus not directly amenable to the technique of multiple linear regression, and has no exact solution. However, it can be solved to a high degree of precision using, for example, a method that linearizes the problem [121].

Figure 7.1A shows the fit of Eq. 7.2 to data obtained during a passive expiration from a dog that was anesthetized, paralyzed, and tracheostomized [123]. The animal's lungs were inflated with a large air-filled syringe and then suddenly allowed to deflate under the influence only of the elastic recoil of the respiratory system (which in this case included the tissues of both the lungs and the chest wall). It is clear from Fig. 7.1A that the fit is not very good. In particular, there are rather substantial systematic deviations between the measured and fitted curves that indicate that the model does not have sufficient degrees of freedom to account for all the systematic variation in the data. The linear single-compartment model would thus appear to have some major shortcomings as a representation of the respiratory system that produced these data.

So, if a single exponential does not give a good fit, what about two exponentials at the same time? That is,

$$V(t) = V_1 e^{-t/\tau_1} + V_2 e^{-t/\tau_2} \qquad (7.4)$$

where $V_1 + V_2 = V_0$. Figure 7.1B shows that this gives an extremely good fit. Indeed, the only deviations between the data and the fit appear to be undulations due to the mechanical interaction between the lungs and the beating heart (so-called cardiogenic oscillations). Clearly, in order to account for a relaxed expiration, we should be looking for a model that predicts lung volume to decrease in a bi-exponential fashion.

7.2   Two-compartment models of heterogeneous ventilation

Early efforts to understand mechanical heterogeneity in the lung began decades ago with attempts to explain the frequency dependence of E [124, 125]. At that time, E and R had to be determined using special maneuvers involving oscilloscopes or graph paper, and usually focused on only a few key data points while discarding the rest (Section 3.2.4). Digital computers have now supplanted such methods with more robust and convenient numerical techniques (Section 3.2.1). Nevertheless, the older methods still allowed investigators to discover that E increases with breathing frequency. Actually, the original studies focused on compliance, the inverse of E, and noticed that it decreased with frequency. This phenomenon came to be known as the *frequency dependence of compliance* [124].

The single-compartment linear model of the lung predicts that the values of the parameters E and R should be independent of the frequency at which flow (or volume) is oscillated in and out of the lung. An extension of this model is called for if we are to

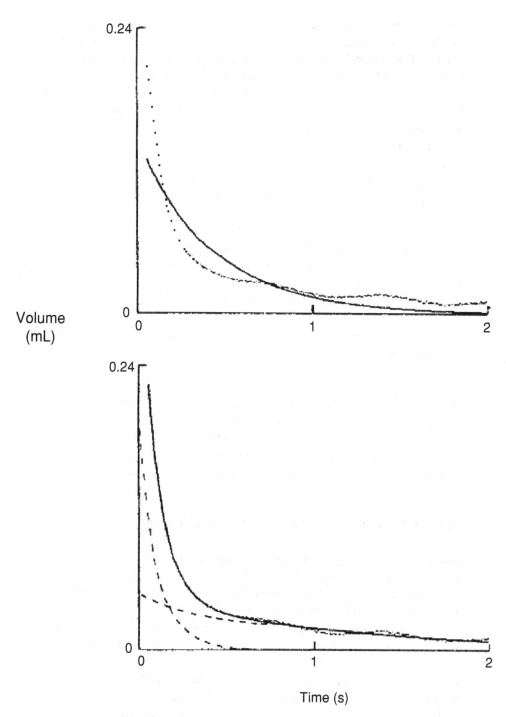

Figure 7.1 (A) Volume-time profile of the lungs recorded during a passive expiration in a dog (dots) and the fit provided by a single exponential function (solid line). (B) The same data with a double-exponential fit. The two exponentials are shown individually as dashed lines (adapted with permission from [122]).

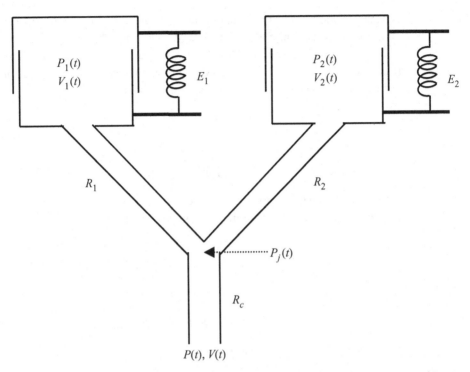

Figure 7.2 Parallel two-compartment model. The model parameters are R_1, E_1, R_2, E_2, and R_c. The compartmental pressures and volumes are $P_1(t)$, $V_1(t)$, and $P_2(t)$, $V_2(t)$, respectively. Pressure and volume at the airway opening are $P(t)$ and $V(t)$, respectively.

understand both the genesis of frequency dependence and the double-exponential decay of lung volume during a relaxed expiration.

At the next level of complexity up from the single-compartment model there are, obviously, models with two compartments. We say that such a model has two mechanical degrees of freedom because the state of the model at any point in time is uniquely specified by two quantities, namely the volumes of the two compartments. As we shall see, two-compartment models predict relaxed expired volume to proceed according to a bi-exponential function of time in agreement with observation (Fig. 7.1B). At this point, however, things begin to get complicated because there is more than one physiologically plausible way to arrange two compartments.

7.2.1   The parallel model

One possible two-compartment model is the parallel arrangement shown in Fig. 7.2, which represents two different alveolar regions, each with its own local airway connected to the airway opening via a common airway. This model (without the common airway) was first investigated in a pulmonary context about 50 years ago [124], and introduced to physiology the notion that parallel ventilation heterogeneity should exist in the lung and that it could have a significant effect on the dynamic nature of lung mechanics. This model has been a strong force in the field of pulmonary physiology ever since.

To determine how this model behaves, we start with the equations that describe relationships between pressures, flows, and volumes in each of its components. The compartmental elastic pressures give the pair of equations

$$P_1(t) = E_1 V_1(t) \tag{7.5}$$

and

$$P_2(t) = E_2 V_2(t). \tag{7.6}$$

To write the expressions for resistive pressure drops along the airways, we define the pressure at the junction of the three airways as $P_j(t)$. This gives

$$P_j(t) - P_1(t) = P_j(t) - E_1 V_1(t)$$
$$= R_1 \dot{V}_1(t) \tag{7.7}$$

and

$$P_j(t) - E_2 V_2(t) = R_2 \dot{V}_2(t). \tag{7.8}$$

The equation for the common airway is

$$P(t) - P_j(t) = R_c[\dot{V}_1(t) + \dot{V}_2(t)]. \tag{7.9}$$

Substituting for $P_j(t)$ from Eq. 7.9 into Eqs. 7.7 and 7.8 gives a pair of coupled first-order ordinary differential equations,

$$P(t) = E_1 V_1(t) + (R_1 + R_c) \dot{V}_1(t) + R_c \dot{V}_2(t) \tag{7.10}$$

and

$$P(t) = E_2 V_2(t) + (R_2 + R_c) \dot{V}_2(t) + R_c \dot{V}_1(t). \tag{7.11}$$

Now, what we are after is an equation that links together the two variables we can measure, namely $P(t)$ and $V(t)$. This means $V_1(t)$, $V_2(t)$, and their time-derivatives have to be eliminated between Eqs. 7.10 and 7.11 and replaced with $V(t)$ and $\dot{V}(t)$. This would seem to pose a difficulty given that with two equations one can eliminate only a single unknown. The trick here, however, is to differentiate Eqs. 7.10 and 7.11 with respect to time to obtain, respectively,

$$\dot{P}(t) = E_1 \dot{V}_1(t) + (R_1 + R_c) \ddot{V}_1(t) + R_c \ddot{V}_2(t) \tag{7.12}$$

and

$$\dot{P}(t) = E_2 \dot{V}_2(t) + (R_2 + R_c) \ddot{V}_2(t) + R_c \ddot{V}_1(t). \tag{7.13}$$

Now substitute for $\ddot{V}_2(t)$ from Eq. 7.12 into Eq. 7.13 to obtain

$$R_2 \dot{P}(t) = (R_2 + R_c) E_1 \dot{V}_1(t) + (R_1 R_2 + R_1 R_c + R_2 R_c) \ddot{V}_1(t) - R_c E_2 \dot{V}_2(t). \tag{7.14}$$

Next, replace $\dot{V}_2(t)$ in Eq. 7.14 with its expression derived from Eq. 7.10 to obtain an equation for compartment 1 alone as

$$R_2 \dot{P}(t) + E_2 P(t) = [R_1 R_2 + R_c (R_1 + R_2)] \ddot{V}_1(t)$$
$$+ [(R_2 + R_c) E_1 + (R_1 + R_c) E_2] \dot{V}_1(t) + E_1 E_2 V_1. \quad (7.15)$$

Because of the symmetry of the parallel model (Fig. 7.2), a corresponding equation can be written for compartment 2 by interchanging subscripts 1 and 2. Noting that $V(t) = V_1(t) + V_2(t)$, the equation for compartment 2 can be added to Eq. 7.15 to give the equation for the entire model. This equation is

$$(R_1 + R_2) \dot{P}(t) + (E_1 + E_2) P(t) = [R_1 R_2 + R_c (R_1 + R_2)] \ddot{V}(t)$$
$$+ [(R_2 + R_c) E_1 + (R_1 + R_c) E_2] \dot{V}(t)$$
$$+ E_1 E_2 V. \quad (7.16)$$

Note that interchanging subscripts 1 and 2 leaves both the above equation and the model in Fig. 7.2 unchanged.

We thus find the equation of motion of the parallel two-compartment model to be a *second-order linear differential equation*. This process can be repeated, in principle, for models with more compartments, and the pattern should by now be clear; a three-compartment model is governed by a third-order differential equation, and so on. However, the reader who journeyed through the above derivation should also be convinced that attempting this process for a model of more than two compartments is not for the algebraically faint of heart.

An alternative (and more standard) approach to finding the equation for $V(t)$ begins by taking the *Laplace transforms* of Eqs. 7.12 and 7.13, as any university-level text on differential equations will show. This reduces them from a system of coupled linear differential equations to a system of simultaneous algebraic equations, making the process of finding the equation for $V(t)$ more tractable. Nevertheless, the algebra involved still increases in complexity very rapidly as the number of compartments increases.

The *homogeneous* form of Eq. 7.16 is obtained by setting $P(t)$ and its derivative equal to zero. This is the equation that describes how the model behaves during a relaxed expiration when pressure at the mouth is zero. That is,

$$[R_1 R_2 + (R_1 + R_2) R_c] \ddot{V}(t) + [E_1 R_2 + E_2 R_1 + (E_1 + E_2) R_c] \dot{V}(t) + E_1 E_2 V(t) = 0. \quad (7.17)$$

The solution to Eq. 7.17 is a double-exponential expression of the form

$$V(t) = A_1 e^{-t/\tau_1} + A_2 e^{-t/\tau_2}. \quad (7.18)$$

Substituting Eq. 7.18 into Eq. 7.17 shows that $1/\tau_1$ and $1/\tau_2$ are both solutions to the quadratic equation

$$\frac{1}{\tau^2} + \frac{[E_1 R_2 + E_2 R_1 + (E_1 + E_2) R_c]}{[R_1 R_2 + (R_1 + R_2) R_c]} \frac{1}{\tau} + \frac{E_1 E_2}{[R_1 R_2 + (R_1 + R_2) R_c]} = 0. \quad (7.19)$$

That is,

$$\tau_1, \tau_2 = \frac{-[E_1 R_2 + E_2 R_1 + (E_1 + E_2)R_c]}{2[R_1 R_2 + (R_1 + R_2)R_c]}$$

$$\pm \sqrt{\frac{\dfrac{[E_1 R_2 + E_2 R_1 + (E_1 + E_2)R_c]^2}{[R_1 R_2 + (R_1 + R_2)R_c]^2} - \dfrac{4E_1 E_2}{[R_1 R_2 + (R_1 + R_2)R_c]}}{4}}. \quad (7.20)$$

The initial volume of the model, $V(0)$, is the sum of the two compartments' volumes, each determined by the inflation pressure, P_0, applied just prior to expiration. From Eq. 7.18, this gives

$$V(0) = P_0 \left(\frac{1}{E_1} + \frac{1}{E_2} \right)$$

$$= A_1 + A_2. \quad (7.21)$$

Also, the initial flow from the model is the sum of the compartmental flows, each determined by P_0 and the downstream resistances. That is,

$$\frac{dV(0)}{dt} = P_0 \left(\frac{1}{R_1 + R_c} + \frac{1}{R_2 + R_c} \right)$$

$$= \frac{-A_1}{\tau_1} - \frac{A_2}{\tau_1}, \quad (7.22)$$

from which we obtain

$$A_1 = \frac{\tau_1}{\tau_1 + \tau_2} \left[P_0 \left(\frac{1}{E_1} + \frac{1}{E_2} \right) + \tau_2 P_0 \left(\frac{1}{R_1 + R_c} + \frac{1}{R_2 + R_c} \right) \right] \quad (7.23)$$

and

$$A_2 = \frac{\tau_2}{\tau_1 + \tau_2} \left[P_0 \left(\frac{1}{E_1} + \frac{1}{E_2} \right) + \tau_1 P_0 \left(\frac{1}{R_1 + R_c} + \frac{1}{R_2 + R_c} \right) \right]. \quad (7.24)$$

Again, a useful check on the correctness of the algebra is to note that interchanging subscripts 1 and 2 in Eq. 7.23 produces Eq. 7.24, and vice versa, as must be the case given the symmetry of the model.

7.2.2   The series model

An alternative arrangement for two compartments is to connect them in series [125], as might represent the distal parenchyma and the proximal airways, respectively (Fig. 7.3). The equation of motion of the series model can be derived in an analogous way to that given above for the parallel model. Actually, the equation for the series model is easier to derive because there is no junction pressure $P_j(t)$ involved. The two coupled first-order differential equations describing the behavior of each compartment are

$$P(t) = E_1 V_1(t) + R_1 \dot{V}(t) \quad (7.25)$$

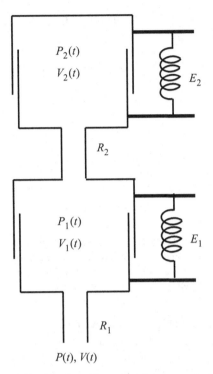

Figure 7.3 The series two-compartment model of the lung. The model parameters are R_1, E_1, R_2, and E_2. The compartmental pressures and volumes are $P_1(t)$, $V_1(t)$ and $P_2(t)$, $V_2(t)$, respectively. Pressure and volume at the airway opening, $P(t)$ and $V(t)$ respectively, are matched to experimental measurements.

and

$$E_1 V_1(t) - E_2(V(t) - V_1(t)) = R_2(\dot{V}(t) - \dot{V}_2(t)),\tag{7.26}$$

which leads to

$$R_2 \dot{P}(t) + (E_1 + E_2)P(t) = R_1 R_2 \ddot{V}(t) + (E_1 R_2 + E_2 R_1 + R_1 E_1)\dot{V}(t) + E_1 E_2 V(t).\tag{7.27}$$

Note that Eq. 7.27 is not symmetric in the indices 1 and 2, as befits the model in Fig. 7.3, which is also not symmetric.

Both the parallel and series models thus have equations of motion of the form

$$\dot{P}(t) + A_1 P(t) = A_2 \ddot{V}(t) + A_3 \dot{V}(t) + A_4 V(t)\tag{7.28}$$

where the A_i are determined by the values of the compartment elastances and airway resistances in each model as per Eqs. 7.16 and 7.27, respectively. This means that both models predict the change in lung volume during a relaxed expiration to be a double-exponential function of time (Eq. 7.18). Furthermore, the parameters of this function, V_1, V_2, τ_1, and τ_2, can be related to the parameters of either model with equal

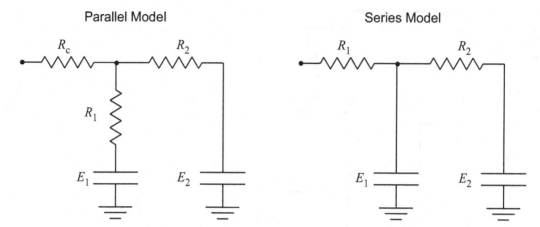

Figure 7.4 Electrical circuit analogs of the parallel and series compartment lung models shown in Figs. 7.2 and 7.3, respectively. The various parameters of the electrical models mirror the corresponding parameters in the compartment models. (Note that the usual parameter associated with an electrical capacitor is capacitance, which is the inverse of elastance.)

ease mathematically. It is therefore impossible to decide which of the two models is appropriate for modeling the lung on the basis of measurements of $P(t)$ and $V(t)$ alone. This is an example of what is known as a *model uniqueness problem*; there is more than one physiologically plausible model structure that can account equally well for a given set of experimental data. In fact, one could imagine a third candidate model consisting of a combination of the series and parallel models, realized by including an additional airway laterally between the two compartments in Fig. 7.2. This pathway would presumably represent the collateral pathways which have been identified in humans and several other species [126], and which are thought to aid in the redistribution of flow throughout the lung periphery.

7.2.3   Electrical analogs

The models shown in Figs. 7.2 and 7.3 represent physical analogs of the lung because they are characterized by the same variables, namely pressure, volume, and flow of gas. It is common practice, however, to represent lung models in terms of *electrical analogs* in which the mathematics involved is congruent to that of the physical models even though the variables are different. In electrical circuit models, pressure is replaced by *voltage*, volume by *charge*, and flow by *current*. Flow-resistive airways are replaced by *electrical resistors*, while elastic compartments are replaced by *capacitors*.

Figure 7.4 shows the electrical circuit representations of the parallel and series compartment models. The equations describing the electrical models in Fig. 7.4 are determined by first expressing the relationships between voltage, current, and charge in each of their arms. This yields pairs of coupled first-order differential equations that can be combined, using the same methods as described above for the compartment models, to produce single equations relating total current through each model to the voltage across it. These equations are exactly the same as those for the compartment models (Eqs. 7.16 and 7.27) if pressure is replaced by voltage and flow by current. Such mathematical congruence defines the models as being precisely equivalent.

It can be confusing to move between physical and electrical models of the lung because their structures on paper do not always look the same, as illustrated by the lung models introduced in this chapter. The naming of these models as "parallel" and "series" might seem obvious from inspection of their physical representations in Figs. 7.2 and 7.3, respectively. However, their electrical representations (Fig. 7.4) belie this interpretation, as both appear to be manifestly parallel. The key to avoiding confusion on this issue is to determine whether flow (or current) is either experienced equally by both compartments (capacitors) or shared between them. In the case of the parallel model in Fig. 7.2, it is obvious that the total flow into the model splits at the end of the common airway into two components that each go to a different compartment. However, the same is also true of the series model in Fig. 7.3. That is, a portion of the total flow entering through R_1 is retained in the proximal compartment causing it to inflate, while the remainder passes through to the distal compartment. Thus, from a mathematical point of view, both models are actually parallel. The parallel and series designations conventionally assigned to these models in the pulmonary physiology literature are thus reflective only of their anatomical interpretations.

7.3   A model of tissue viscoelasticity

When two lung models are governed by a differential equation of the same form (such as Eq. 7.28), they are indistinguishable in their abilities to mimic overall lung behavior. Thus, for any choice of values for the parameters of the parallel model in Fig. 7.2, there is a set of parameters for the series model in Fig. 7.3 that will predict exactly the same relationships between $P(t)$, $V(t)$, and $\dot{V}(t)$. On the other hand, the physiological interpretations of these two models are entirely different. To make matters worse, there is yet another physiologically plausible model which is potentially appropriate for describing the two-compartment behavior of the lung. This alternative model, which actually pre-dates the parallel and series models [127], consists of only a single homogeneously ventilated alveolar compartment, but the tissue comprising the compartment is *viscoelastic*. This means that the pressure inside the compartment is not simply a function of its volume, but also depends on its volume history.

The *viscoelastic model* is shown in Figure 7.5 and illustrates why we have been drawing the compartments in the previous models as telescoping canisters coupled by springs instead of the more conventional elastic balloons. It is easy to incorporate additional elements linking the canisters to make the mechanical coupling between them more complicated. The single conduit of the viscoelastic model represents the entire conducting airway tree and therefore has resistance R_{aw}. The mechanical properties of the tissues, however, are now represented by three elements: a resistor (known as a *dashpot*) and two springs. The three elements R_t, E_1, and E_2 in Fig. 7.5 together constitute what is known as a *Kelvin body*. The stiffness of the spring E_1 represents the

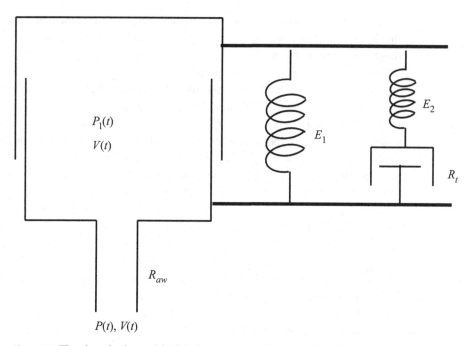

Figure 7.5 The viscoelastic model of the lung with two degrees of freedom. The alveolar regions are represented as a single uniformly ventilated compartment served by an airway with resistance R_{aw}. The lung tissues are described by a Kelvin body consisting of three elements, E_1, E_2, and R_t.

static elastic behavior of the lung, while the series combination of R_t and E_2 (which together constitute a *Maxwell body*) account for its viscoelastic behavior.

The model in Fig. 7.5 represents only a single physical compartment in the lung, but it still has two degrees of freedom because its state at any point in time is defined by two quantities, the volume in the alveolar compartment and the extension of the spring E_2 (or, equivalently, the extension of the dashpot). Let the extension of the spring E_2 be $x(t)$. The two coupled first-order differential equations describing $V(t)$ and $x(t)$ are then

$$P(t) = E_1 V(t) + E_2 x(t) + R_{aw} \dot{V}(t) \qquad (7.29)$$

and

$$E_2 x(t) = R_t(\dot{V}(t) - \dot{x}(t)). \qquad (7.30)$$

The resulting equation of motion for the complete viscoelastic model is

$$R_t \dot{P}(t) + E_2 P(t) = R_{aw} R_t \ddot{V}(t) + (E_1 R_t + E_2 R_{aw} + E_2 R_t) V(t) + E_1 E_2 V(t), \qquad (7.31)$$

which again has the form of Eq. 7.28. Thus, although this is only a single-compartment model in terms of distinct ventilation units, it ranks as another example of a two-compartment model from a mathematical perspective because it is described by a second-order differential equation.

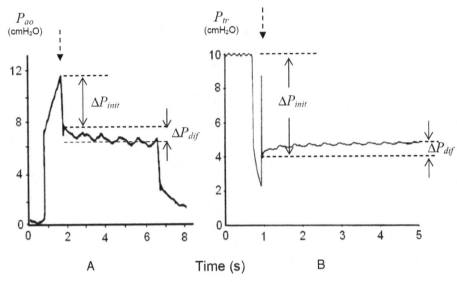

Figure 7.6 (A) Example of stress relaxation in airway opening pressure (P_{ao}) measured in a human subject following a sudden interruption of inspiratory flow (reproduced with permission from [130]). (B) Stress recovery in tracheal pressure (P_{tr}) measured in an anesthetized, tracheostomized dog following interruption of flow during a passive expiration (reproduced with permission from [131]). The vertical dashed arrows indicate the instants of flow interruption. ΔP_{init} is the initial change in pressure that occurs immediately upon interruption of flow, while ΔP_{dif} is the further transient change that occurs as a result of stress recovery in the lung tissue. Note that the pressure tracings from both human and dog exhibit marked cardiogenic oscillations produced by the heart impacting the lung with each beat.

7.4   Stress adaptation and frequency dependence

All the two-compartment models we have considered above are potentially capable of explaining the key experimental phenomena that caused investigators to look beyond the single-compartment model as a representation of lung mechanics. All models, for example, exhibit *stress adaptation*. This is observed experimentally as a transient change in pressure at the airway opening following sudden interruption of flow leaving or entering the lungs. When flow is interrupted, the pressure measured just behind the point of interruption equals the pressure deep within the lung, because without any flow there can be no pressure gradient along the airways. Interruption of flow during inspiration (Fig. 7.6A) typically produces an initial drop in pressure (ΔP_{init}) followed by a further slower drop (ΔP_{dif}). ΔP_{init} reflects the immediate obliteration of the resistive pressure drop along the airways, while ΔP_{dif} is due to the phenomenon known as *stress relaxation*. Interruption of flow during expiration produces corresponding pressure changes in reverse (Fig. 7.6B), with ΔP_{dif} now reflecting *stress recovery*.

The parallel and series models attribute ΔP_{dif} in Fig. 7.6 to redistribution of gas between compartments [13, 128, 129]. Gas redistribution, also known as *pendelluft*, occurs because the pressures P_1 and P_2 within the two model compartments

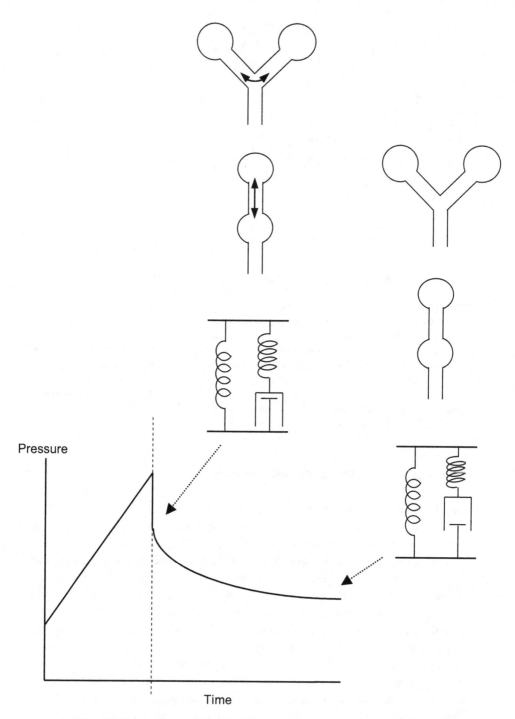

Figure 7.7 Schematic representation of airway opening pressure when flow is suddenly interrupted during inspiration (dashed vertical line). The subsequent pressure relaxation can be ascribed either to redistribution of flow between the two compartments of the parallel or series model, or to relaxation of stress in the spring of a Maxwell body via sliding of its associated dashpot.

(Figs. 7.2 and 7.3) are usually different at the instant of flow interruption. Indeed, if these pressures were not different at least some of the time then these models would behave simply like a model with a single compartment of volume $V_1 + V_2$. In the case of the parallel model, the two compartments have different time-constants of filling and emptying so they gain and lose pressure at different rates during ventilation. In the case of the series model, the only way that gas can move from the proximal to the distal compartment is if the two compartments are at different pressures, so this must pertain most of the time. When flow into the model stops, the two compartmental pressures are free to equilibrate through the flow of an appropriate volume of gas from the higher-pressure compartment to the lower-pressure compartment. This causes the two compartmental pressures to asymptote to an intermediate common pressure, which is reflected in $P(t)$ (Fig. 7.7). The two compartmental pressures are not generally equal at the instant of flow interruption in the series model (Fig. 7.3) either, so again there is an exchange of gas along the airway R_2 to eliminate this pressure difference. This produces a transient change in $P_1(t)$ which is reflected in $P(t)$ at the airway opening [128].

The way that stress adaptation occurs in the viscoelastic model can be appreciated by another thought experiment. Imagine that the model in Fig. 7.5 is rapidly inflated and then maintained at the elevated volume. Immediately following the inflation, both springs E_1 and E_2 will be stretched, so the pressure in the alveolar compartment will reflect the sum of their tensions. Subsequently, however, spring E_2 will relax by moving the dashpot R_t. This will cause alveolar pressure to decrease transiently to the level maintained only by spring E_1. The result is stress relaxation in $P(t)$ (Fig. 7.6A). The reverse process, *stress recovery*, occurs if volume is suddenly removed from the model (Fig 7.6B).

The frequency dependence of resistance and elastance is another phenomenon seen in real lungs that cannot be reproduced by the single-compartment model. If lung volume is oscillated sinusoidally and the resulting $P(t)$ and $\dot{V}(t)$ signals measured at the airway opening are fit to the single-compartment model, one obtains values of R and E that depend on the frequency of oscillation. In particular, R decreases monotonically with frequency while E increases. Because all the two-compartment models considered above are described by a differential equation of the same form, Eq. 7.28, they have equal ability to reproduce this behavior.

The mechanism by which frequency dependence occurs is most easily seen with respect to the viscoelastic model (Fig. 7.5). When flow oscillates into the alveolar compartment of this model very slowly, the dashpot R_t is given sufficient time to move under the influence of E_2. That is, as soon as E_2 becomes stretched by even a little bit, it exerts a force on R_t, which then slides toward the force in order to eliminate it. Consequently, the elastic behavior of the system is determined almost entirely by E_1, which always follows the alveolar volume changes precisely. By the same token, the system dissipates energy through R_t, so the tissues exhibit resistance. Conversely, when alveolar volume changes occur at a high frequency, the dashpot is given little time to slide in response to any tension in E_2 before the tension reverses direction. Consequently, the spring E_2 is forced to follow the alveolar volume changes, so the elastic properties of the system are given by the sum of both E_1 and E_2, with minimal dissipation occurring in R_t. As frequency increases, the apparent tissue resistance proceeds toward zero.

7.5   Resolving the model ambiguity problem

So far we have seen that the normal lung behaves a lot more like a system of two linear compartments than of only one, at least during maneuvers like relaxed expiration, flow interruption, and sinusoidal oscillation at different frequencies. We have also seen that this behavior can potentially be explained by either parallel or series differences in ventilation time-constants, or by tissue viscoelasticity. Furthermore, we cannot distinguish between these various models from measurements of $P(t)$, $V(t)$, and $\dot{V}(t)$ made at the airway opening, regardless of what experimental maneuvers are performed, because the differential equation that relates these variables has precisely the same form for all the models (Eq. 7.28). Viewed simply from the perspective of $P(t)$, $V(t)$, and $\dot{V}(t)$, it does not matter which of these mechanisms is operative. It matters a great deal, however, from the perspective of physiological insight. Resolving this model ambiguity problem is therefore an important issue.

The only way to resolve a model ambiguity problem of this nature is to perform a new kind of experiment, one that can distinguish between the inner workings of the different models. This means measuring at least one of the individual state variables. In the case of the ventilation heterogeneity models, these state variables are $V_1(t)$ and $V_2(t)$. For the viscoelastic model, the variables are $V_1(t)$ and $x(t)$.

The experimental approach that was used to resolve this particular model ambiguity problem employed alveolar capsules to measure alveolar pressure, $P_A(t)$, at three different sites on the lung surface while sudden interruptions of flow were performed at the airway opening during passive expiration. These experiments were performed in normal, anesthetized, tracheostomized dogs [131], and Fig. 7.8 shows a typical example of the data obtained. When the lungs were held inflated at a pressure of about 8 cmH$_2$O, the pressure at the entrance to the trachea (P_{tr}) and three different alveolar capsule measurements of P_A were all identical. Then, when expiration was allowed to proceed, P_{tr} dropped immediately to a level reflecting the small resistive pressure drop across the measuring equipment. The three P_A signals, on the other hand, measured the decreasing elastic recoil of the lung tissue and so descended more gradually. Importantly, however, these three P_A were essentially identical throughout expiration, indicating that the lungs emptied homogeneously.

The most critical part of this experiment occurred upon sudden interruption of flow, which took place about 200 ms into expiration. Here, the P_A signals all stopped decreasing and started on the same slow recovery toward an eventual plateau (the plateau was reached after about 5 s). Most importantly, however, P_{tr} joined the P_A signals on their common post-interruption trajectory. To do this, P_{tr} had to make a sudden jump up from its low pre-interruption level to reach the level of the three P_A signals. Actually, P_{tr} overshot this level briefly (see the spike in P_{tr} just under the vertical dashed arrow in Fig. 7.8). This overshoot was due to high-frequency axial oscillations of the gas in the

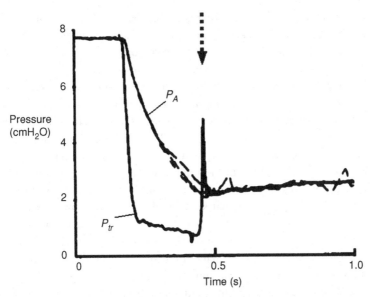

Figure 7.8 Pressures measured during expiratory flow interruption in an anesthetized, tracheostomized dog. Tracheal pressure (P_{tr}) and three different alveolar pressures (P_A) began at the same point prior to the beginning of expiration. They came together again immediately after the interruption of flow. The instant of flow interruption is shown by the vertical dotted arrow (adapted with permission from [131]).

central airways [132], but these were very quickly damped out. These oscillations aside, one can see in Fig. 7.8 that the sudden change in P_{tr} upon flow interruption was equal to the resistive pressure drop across the airways that existed at the instant of interruption. This sudden pressure change is also apparent in Fig. 7.6B as ΔP_{init}, and is seen in reverse in Fig. 7.6A. Dividing ΔP_{init} by the flow measured immediately prior to interruption thus gives a measure of airway resistance, at least in an open-chest dog [131]. The subsequent slow recovery in pressure, labeled ΔP_{dif} in Fig. 7.6, is common to P_{tr} and P_A alike and must therefore have been due to stress recovery occurring within the lung tissues.

The data in Fig. 7.8 were obtained with the rib cage open and widely retracted, so they pertain to the lungs alone. When the chest wall is intact, ΔP_{init} contains a contribution from the tissues of the chest wall [17]. Regardless, these alveolar capsule experiments demonstrate that the normal lung behaves like a single homogeneously ventilated alveolar compartment made of viscoelastic tissue that is connected directly to the airway opening. In other words, of the three two-degree-of-freedom models we have examined in this chapter, the viscoelastic model shown in Fig. 7.5 is the most appropriate for describing the normal lung. Most of the normal lung's departure from single-compartment behavior is thus due to the viscoelastic properties of the parenchymal tissues rather than regional variations in ventilation.

Other studies have corroborated the findings of the flow interruption experiments. For example, when the normal lung is mechanically ventilated at frequencies in the range of normal breathing, multiple alveolar capsules located at various sites on the pleural surface show P_A to be remarkably uniform throughout the lung [15, 131]. Of course, it is important to realize that this does not mean the viscoelastic model is *the correct model* of the lung. All we have shown here is that among the possible contenders having two mechanical degrees of freedom, the viscoelastic model is the most appropriate for describing the normal lung following interruption of flow or during oscillation within the breathing frequency range. It is easy to find situations in which this model fails rather badly. One such example is when volume is oscillated at the airway opening at frequencies above the range of normal breathing, which extends up to about 2 Hz. Even by 6 Hz the differences in the time-constants of different lung regions lead to significant differences in P_A [12]. The mechanical behavior of the lung also gets a lot more complex in situations that simulate disease. For example, when a bronchial agonist such as histamine is administered to the lungs to cause the airway smooth muscle to constrict and thus simulate an asthma attack, P_A measured at various sites on the lung surface can become markedly disparate even during normal mechanical ventilation [133, 134]. Under these conditions, the lung no longer behaves like a single uniformly ventilated compartment. Modeling the lung under such conditions would presumably require a construct with multiple viscoelastic compartments. Unfortunately, as compartments are added, the model ambiguity problem grows exponentially. Also, additional compartments mean additional free parameters that must be evaluated from the experimental data, and the more parameters one has to evaluate from a given data set, the greater the uncertainty in the estimated values. For these reasons, inverse models of physiology are invariably rather simple in structure and characterized by only a small number of free parameters.

7.6   Fitting the two-compartment model to data

The reasons why the model ambiguity problem arose with respect to the parallel, series, and viscoelastic models examined above is that they are all described by an equation of exactly the same form, namely Eq. 7.28. Rearranging this equation gives

$$P(t) = \frac{A_2}{A_1}\ddot{V}(t) + \frac{A_3}{A_1}\dot{V}(t) + \frac{A_4}{A_1}V(t) - \frac{1}{A_1}\dot{P}(t) + P_0$$
$$= B_1\ddot{V}(t) + B_2\dot{V}(t) + B_3V(t) + B_4\dot{P}(t) + P_0 \qquad (7.32)$$

where the ubiquitous P_0 term has been added for the reasons outlined in Section 3.1.2. This equation should by now have a familiar look to it. The dependent variable $P(t)$ is predicted to be a sum of five independent variables each scaled by a multiplicative parameter. Consequently, these parameters can all be estimated by multiple linear regression.

By fitting Eq. 7.32 to measurements of P, $\dot{V}$, and V, we can evaluate B_1 through B_4 plus P_0. These parameter values can then be used to calculate the parameters of any of the two-compartment models considered above via their respective equations of motion (Eqs. 7.16, 7.27, and 7.31). There is a practical issue that arises, however. The primary

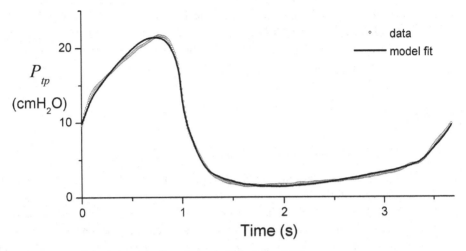

Figure 7.9 Fit of Eq. 7.33 to the data shown in Figs. 5.2 and 5.3.

data consist of measurements of P and $\dot{V}$. To generate the independent variables for Eq. 7.32, $\dot{V}$ not only has to be integrated with respect to time to produce V, but both P and $\dot{V}$ must also be differentiated to produce $\dot{P}$ and $\ddot{V}$, respectively. Differentiation is a noise-amplifying procedure, so this will exacerbate the parameter bias produced by noise in the independent variables (Section 3.2.2). It may even amplify the noise to the point of causing nonsensical parameter estimates to be produced.

A way of ameliorating the problems caused by differentiating noisy data is to recast the fitting problem in terms of the time-integral of Eq. 7.32. That is,

$$P(t) = A_2 \dot{V}(t) + A_3 V(t) + A_4 \int_0^t V(t)\,dt - A_1 \int_0^t P(t)\,dt + A_1 P_0 t + C. \quad (7.33)$$

This introduces an additional parameter that must be estimated, the constant of integration C, but the trade-off is that the noise in the independent variables is greatly reduced [135].

Figure 7.9 shows the result of fitting Eq. 7.33 to the data that never quite received a satisfactory fit from the nonlinear single-compartment models in Chapter 5. Compared to the fits achieved in Figs. 5.2 and 5.3, the fit in Fig. 7.9 is spectacular, with a CD of 0.998 and an $AICc$ value of 151. This is a huge reduction in $AICc$ compared to the values of well over 1000 produced with the same data by the linear single-compartment model and its two nonlinear extensions (Section 5.3.2). In other words, the lung that produced these data behaved much more like a system of two linear compartments than a single compartment with nonlinear properties.

The values of A_1, A_2, A_3, and A_4 estimated via Eq. 7.33 can now be converted into values for the parameters of a two-compartment model by equating terms in Eq. 7.33 to either of Eqs. 7.16, 7.27, or 7.31. Which model it should be remains uncertain, however. The most appropriate choice is likely to be the viscoelastic model in Fig. 7.5 because this is the best model for describing the lungs of a normal dog [135]. However, the data in

Fig. 7.9 were collected in a human patient receiving mechanical ventilation for respiratory failure, so we cannot rule out the possibility that ventilation heterogeneity may have been significant. Without some independent means of verifying this, it is impossible to be sure either way.

Problems

7.1   Demonstrate that Eq. 7.18 is a solution to Eq. 7.17.

7.2   Volume decays in a single exponential fashion during a passive expiration if the lung behaves like a single compartment with resistance and elastance values that remain fixed. How would the time-course of expiration change if the resistance depended on flow as in Eq. 5.1? What about if elastance depended on volume as in Eq. 5.3? Suppose that P_{el} in Eq. 5.3 was equal only to $E_2 V^2$ while resistance remained constant. What would the functional form for a relaxed expiration be?

7.3   Derive Eq. 7.27 from Eqs. 7.25 and 7.26.

7.4   Draw the electrical circuit analog of the viscoelastic model shown in Fig. 7.5.

7.5   Add a second Maxwell body (spring and dashpot) next to the one that is already there in the viscoelastic model shown in Fig. 7.5. Write down the set of three coupled first-order differential equations that describes this extended model. What functional form will stress adaptation take in this model following a sudden interruption of expiratory flow?

7.6   Relate the coefficients of Eq. 7.33 to the parameters of the viscoelastic model shown in Fig. 7.5.

7.7   Following sudden interruption of air flow at the mouth, there is an immediate change in mouth pressure (ΔP_{init} in Fig. 7.6A). What does this pressure correspond to in terms of the parallel two-compartment model (Fig. 7.2), the series model (Fig. 7.3), and the viscoelastic model (Fig. 7.5)?

8 The general linear model

Two-compartment models do a good job of accounting for the behavior of the normal lung over a modest range of ventilation frequencies or stress-adaptation time scales. Of course, these models do not come close to representing the structural complexities of a real lung, so one can easily imagine that a model with more than two compartments might provide an improved account of lung mechanical behavior. This is particularly true for behavior pertaining to extended scales of time or frequency, or when the lungs become heterogeneous in disease. In principle, there is no limit to the number of compartments such a model could possess. Dealing with such models might sound like a daunting prospect, given the algebraic machinations presented in the previous chapter for models with only two compartments. Fortunately, the tools of linear systems theory, and the *fast Fourier transform* (FFT) in particular, come to the rescue. These tools apply so long as the lung can be considered to behave as a collection of linear compartments, each behaving like the model in Fig. 3.1. In this chapter, we examine the principal tools of linear systems theory and see how they apply to the study of lung mechanics.

8.1 Linear systems theory

A *system* is a set of components that have some kind of collective identity. Systems interact with their environments by receiving *inputs* and producing *outputs*. The relationships between these inputs and outputs are determined by processes within the system. These processes are generally not directly accessible to an outside observer. Nevertheless, the observer may investigate the system indirectly by subjecting it to controlled inputs and observing the outputs that are produced as a result. To put this in pulmonary terms, the lung is conventionally viewed as a single-input, single-output system, with flow applied to the trachea as the input and pressure measured at the tracheal entrance as the output. The reason for this convention will become apparent when we discuss impedance, but there is nothing fundamental about it. Other definitions are possible. For example, tracheal pressure could just as easily be the input variable and flow the output. Inputs and outputs can also be defined at other sites, such as flow at the body surface or pressure in the pleural space.

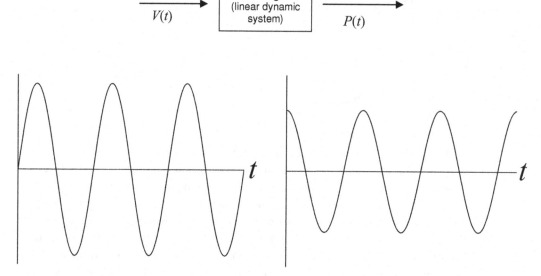

Figure 8.1 Block diagram representation of the lung as a linear dynamic system. An input sine wave of a given frequency will be transformed into an output sine wave of the same frequency, but generally with a different amplitude and phase.

8.1.1   Linear dynamic systems

A *dynamic system* is one in which time plays a role. The lung is obviously a dynamic system because of the nature of its input-output behavior. That is, when a flow, $\dot{V}(t)$, is suddenly applied to the lungs, the resulting effects on pressure, $P(t)$, are not all manifest immediately; some effects take time to develop. If the lung behaves like a *linear dynamic system* then its behavior is described by a linear differential equation that contains terms in the derivatives of $V(t)$ and $P(t)$ up to arbitrary order. We have already encountered such equations of second order in Chapter 7. Extrapolating from these equations, in particular Eq. 7.28, it is easy to see that the general nth-order linear differential equation, which describes a model with n compartments, has the form

$$\sum_{k=0}^{n-1} b_k \frac{d^k P(t)}{dt^k} = \sum_{k=0}^{n} a_k \frac{d^k V(t)}{dt^k} \tag{8.1}$$

where the a_k and b_k are constants and we have omitted the pressure offset term P_0. (Note that the derivative of order zero of a function is simply the function itself.)

There is, in principle, no restriction on the kind of input signals one could apply to a linear dynamic system, practical considerations aside. However, there is one kind of input that is special because it does not change its shape in being transformed from input to output. This special input is the sine wave (Fig. 8.1). To see why this particular functional form enjoys special status, suppose that $V(t)$ is a sine wave with frequency f

and unity amplitude, expressed as

$$V(t) = \sin(2\pi f t). \tag{8.2}$$

The time-derivative of $V(t)$ is

$$\frac{dV(t)}{dt} = 2\pi f \cos(2\pi f t). \tag{8.3}$$

The time-derivative of Eq. 8.3 is

$$\frac{d^2 V(t)}{dt^2} = -4\pi^2 f^2 \sin(2\pi f t). \tag{8.4}$$

Repeated differentiation of a sinusoidal $V(t)$ thus produces terms containing either $\cos(2\pi f t)$ or $\sin(2\pi f t)$ depending on whether the number of differentiations is odd or even. When Eq. 8.2 is substituted for $V(t)$ in Eq. 8.1 the result is

$$\sum_{k=0}^{n-1} b_k \frac{d^k P(t)}{dt^k} = \sum_{k=0}^{n} a_k \frac{d^k \sin(2\pi f t)}{dt^k}$$

$$= \sum_{\substack{k \leq n \\ k\ even}} (-1)^{k/2} (2\pi f)^k a_k \sin(2\pi f t)$$

$$+ \sum_{\substack{k \leq n \\ k\ odd}} (-1)^{(k-1)/2} (2\pi f)^k a_k \cos(2\pi f t). \tag{8.5}$$

Now, the only way that a sum of terms in the time-derivatives of $P(t)$ (i.e. the left-hand side of Eq. 8.5) can equal a sum of sine and cosine terms (i.e. the right-hand side of Eq. 8.5) is if $P(t)$ itself is composed of only sine and cosine terms. That is,

$$P(t) = \alpha \sin(2\pi f t) + \beta \cos(2\pi f t) \tag{8.6}$$

where α and β are constants determined by the various constants on the right-hand side of Eq. 8.5. Using a standard trigonometric identity, Eq. 8.6 can be converted to

$$P(t) = A \sin(2\pi f t + \phi) \tag{8.7}$$

where $A = \sqrt{\alpha^2 + \beta^2}$ and $\phi = \tan^{-1}(\beta/\alpha)$.

Thus, when the input to a linear dynamic system is a sine wave with frequency f, the output will also be a sine wave at the same frequency. In general, however, the output sine wave will be scaled in amplitude by some factor (A in Eq. 8.7) and shifted in phase by some amount (ϕ in Eq. 8.7). Most importantly, while the values of A and ϕ depend on f, they do not depend on the amplitude of the input sine wave. In other words, the shape of a sine wave is preserved when it passes through a linear dynamic system. This shape-preserving property does not apply to any other kind of input.

The fact that A and ϕ depend only on f is of enormous practical importance because it means that a linear dynamic system is completely characterized by the way in which it responds to sine waves of different frequencies. In other words, if you know how A and ϕ depend on f then you know everything there is to know about the system. With this information, and the tools we are about to discuss, you can calculate how the system will respond to any conceivable input. The amplitude and phase functions $A(f)$ and $\phi(f)$ together constitute what is known as the *frequency response* of the system.

8.1.2   Superposition

A linear dynamic system is *time invariant* if its structure and parameter values do not change with time. In other words, the value of n and the a_k and b_k in Eq. 8.1 are fixed. Time invariance never precisely applies to the lung, of course, because its mechanical properties are determined by quantities that change constantly. Airway diameters, for example, vary throughout the breathing cycle and with changes in posture and activity level. Nevertheless, it is possible to make the stationarity assumption over relatively short time periods when conditions are well controlled. Measurements of lung mechanics made under such conditions can then be compared to measurements obtained under different conditions, such as following an experimental intervention.

The advantage of being able to assume that the lung is a *stationary* linear dynamic system is that it then obeys the *principle of superposition*. This means that the way any input $\dot{V}(t)$ is treated by the system is independent of whether or not any other input is present at the same time. For example, suppose that $\dot{V}(t)$ consists of two sine waves having frequencies f_1 and f_2, and that these sine waves can be applied either separately or together. We have already seen that when they are applied separately they will produce sine wave outputs with distinct amplitudes and phases. Superposition means that if the output from the first sine wave is

$$P_1(t) = A_1 \sin(2\pi f t + \phi_1) \tag{8.8}$$

and that from the second one is

$$P_2(t) = A_2 \sin(2\pi f t + \phi_2) \tag{8.9}$$

then the output produced when the two sine waves are applied concurrently is simply

$$P(t) = A_1 \sin(2\pi f t + \phi_1) + A_2 \sin(2\pi f t + \phi_2). \tag{8.10}$$

This can be easily verified by substituting Eqs. 8.8 and 8.9 into Eq. 8.1, which shows that superposition holds in a time-invariant linear dynamic system because differentiation is a linear operation. The principle of superposition applies not just to sine waves, but to any functional form whatsoever, as Fig. 8.2 illustrates.

8.1.3   The impulse and step responses

We have seen above that the sine wave enjoys a place of special significance for linear dynamic systems. There are two other types of input function that also require special mention because of the outputs they produce. These special inputs are the *impulse*, or *Dirac delta function*, and the *step function*.

The delta function, $\delta(t)$, is zero at all times except $t = 0$, where it encompasses an area of unity. $\delta(t)$ is thus a spike of infinite height and infinitesimal width such that the product of its height and width equal 1. Of course, something of infinite height and infinitesimal width cannot be realized in practice, but it can be approximated by a square

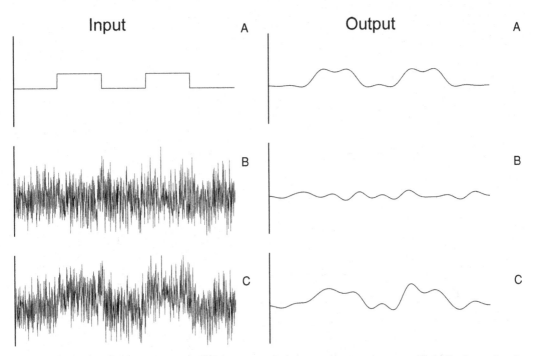

Figure 8.2 The principle of superposition. Panels A and B show two very dissimilar input signals and the outputs that they give rise to. Panel C shows that when these two inputs are added together, the resultant output is simply the sum of the individual outputs.

pulse having a short duration a and height $1/a$ (Fig. 8.3A). The closer a is to zero, the closer the pulse is to a true impulse.

When applied to a linear dynamic system, $\delta(t)$ gives rise to the so-called *impulse response* of the system. For many real-world dynamic systems, including the lung, the impulse response has a finite width. This means that the system is still responding for some time after the impulse itself has come and gone. Therefore, although it is impossible to apply a true impulse of infinitesimal width in practice, a short pulse can still be a satisfactory approximation to an impulse if it is narrow compared to the width of the response it generates, as in the example in Fig. 8.3B.

Now imagine integrating a delta function from $t = -\infty$ to $t = +\infty$. As integration proceeds from left to right, the accumulated area remains at zero until the impulse is encountered at $t = 0$. At this point the integral immediately jumps to 1, where it remains because an impulse has negligible extent and unity area. The integral of $\delta(t)$ is the *Heaviside step function*, H(t) (Fig. 8.3C). When H(t) is applied to a linear dynamic system, the resulting output is the *step response* (Fig. 8.3D). Practically, it is often easier to apply an input that approximates H(t) rather than $\delta(t)$. Since the step response is the time-integral of the impulse response, the impulse response can be obtained by first measuring the step response and then differentiating it with respect to t.

To determine the impulse and step responses of the linear single-compartment model of the lung (Fig. 3.1), imagine what happens to $P(t)$ when the volume of the compartment is zero and $\dot{V}(t)$ is suddenly increased from 0 to 1 at $t = 0$. This will immediately generate

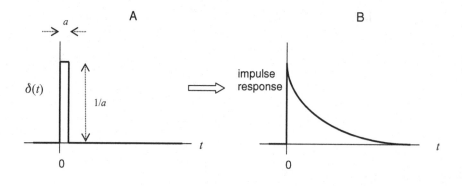

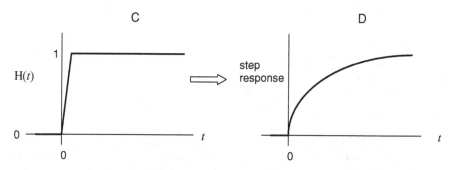

Figure 8.3 (A) An (approximate) impulse $\delta(t)$. (B) The impulse response. (C) The (approximate) Heaviside step function H(t) obtained by integrating the function in A with respect to t. (D) The step response, which is the integral of the function in B.

a fixed resistive pressure drop across the airway equal to $R\dot{V}(t) = R$ as gas flows into the compartment. At the same time, the volume of the compartment will start to increase linearly at a rate of one volume unit per second, so there will be a corresponding linear increase in the elastic recoil pressure within the compartment. The response of the model to a step in $\dot{V}(t)$ will thus be

$$P_{step}(t) = (R + Et)\,H(t) \tag{8.11}$$

where multiplication by H(t) on the right-hand side is required for causality (i.e. P_{step} is non-zero only for $t > 0$). To obtain the impulse response of the model it is now simply a matter of differentiating Eq. 8.11 with respect to t. Note, however, that the right-hand side of Eq. 8.11 consists of the product of two functions of t, so the product rule for differentiation must be invoked. The result is

$$P_{impulse}(t) = (R + Et)\,\delta(t) + EH(t). \tag{8.12}$$

P_{step} and $P_{impulse}$ for the linear single-compartment model are shown in Fig. 8.4. Note that these responses pertain when the input to the model is $\dot{V}(t)$. If we had, for example,

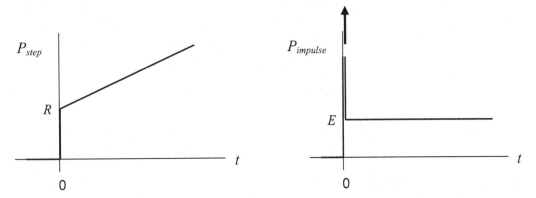

Figure 8.4 The step and impulse responses for the linear single-compartment model of the lung. The vertical arrow above $P_{impulse}$ indicates that the height of the function at this point is infinite while its width is infinitesimal.

been considering $V(t)$ as the input then the step response shown in Fig. 8.4 would be the impulse response.

The response of the lung to a step in flow has received considerable attention in the literature because it describes how the lung behaves during constant-flow inflation. This is a common mode of mechanical ventilation in patients. Consequently, measurements of $P(t)$ and $\dot{V}(t)$ at the trachea have been used to assess respiratory mechanics [130, 136]. The two-compartment models considered in Chapter 7 exhibit a step response of the form shown in Fig. 8.5 [128]. P_{step} exhibits an initial step increase at $t = 0$ due to the sudden establishment of a resistive pressure drop across the airways, just as for the single-compartment model in Fig. 8.4. However, in contrast to the single-compartment model, P_{step} for the two-compartment model exhibits a transient curvilinearity before eventually settling into a steady rate of increase. In the case of the parallel and series models of two compartments (Figs. 7.2 and 7.3, respectively), this transient curvilinearity reflects changing relative flows into the two compartments [128]. In the case of the viscoelastic model (Fig. 7.5), it reflects the development of compression stress in the spring of the Maxwell body that asymptotes to a fixed value as the dashpot approaches constant velocity [136].

8.1.4 Convolution

The importance of the impulse response function for a linear dynamic system stems from the fact that, like the frequency response, it contains all there is to know about the system. That is, if you know the impulse response function, you can calculate how the system will respond to any input. Figure 8.6 shows how this works. An arbitrary input can be approximated by a sequence of narrow adjacent rectangles, each rectangle having a height to match that of the input function. As in Fig. 8.3A, each narrow rectangle approximates a scaled impulse, the scale factor being the area under the rectangle.

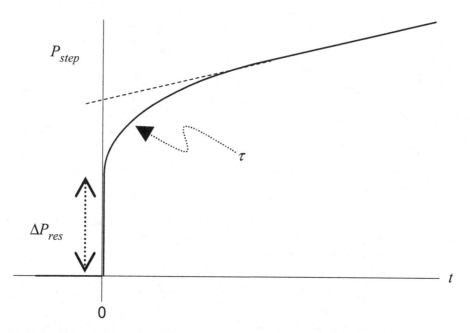

Figure 8.5 Step response of the linear two-compartment model of the lung. The immediate jump ΔP_{res} at $t = 0$ reflects the resistive pressure drop across the airways. The subsequent transient curvature asymptotes to a straight line. The time-constant (τ) of the curved portion reflects the dynamics of either pendelluft between compartments (for either the parallel or series models) or stress adaptation in the tissues (for the viscoelastic model).

Now comes the crucial step. Because the system is linear and obeys the principle of superposition, it treats each impulse independently of all the others. Thus, the sequence of scaled, time-shifted impulses gives rise to a corresponding sequence of scaled and time-shifted impulse responses which add together to produce the complete output, as shown in the bottom panel of Fig. 8.6.

Of course, the scenario depicted in Fig. 8.6 is only an approximation to the truth because a true impulse has infinitesimal width. The picture becomes more accurate as the input rectangles become narrower and more numerous. From a practical point of view, what matters is that the width of each rectangle is narrow compared to the width of the impulse response. If that is the case, then the output calculated as shown in Fig. 8.6 is a good approximation to the true output. Computers allow this calculation to be performed on digitized measurements of $P(t)$ and $\dot{V}(t)$ to any desired degree of accuracy using the formula

$$P_j = \sum_{i=0}^{N} \dot{V}_i h_{j-i} \delta t \tag{8.13}$$

where the h_i are the N digitized samples of the impulse response function and δt is the data sampling interval.

Equation 8.13 can be converted into an exact expression for the output of a linear dynamic system by taking the limit as δt tends to zero. This produces the *convolution*

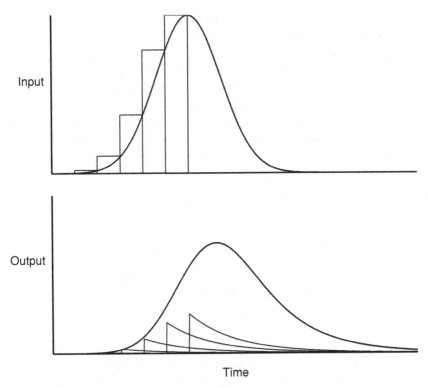

Figure 8.6 A general input can be broken up into a sequence of narrow rectangles approximating scaled impulses (top). The impulses give rise to scaled and time-shifted impulse responses that sum together to produce the complete output (bottom).

integral

$$P(t) = \int_{-\infty}^{\infty} \dot{V}(u)h(t-u)\, du \qquad (8.14)$$

where u is a *dummy variable of integration* that represents time but does not appear in the final output of the calculation. The variable t represents the particular value of time for which the output is determined.

8.2   The Fourier transform

8.2.1   The discrete and fast Fourier transforms

The most important analytical tool in linear systems theory is the Fourier transform, named after the famous French mathematician, Joseph Fourier, who lived around the turn of the nineteenth century. The theory of the Fourier transform was originally developed in terms of continuous functions that involve the use of integrals over time and frequency. However, computer manipulation of biomedical data is always done on finite sets of discrete points, to which the *discrete Fourier transform* applies. We focus here on the discrete theory, which can be understood by appealing to the notion of curve fitting, as follows.

Suppose we have two measurements of pressure made at two different times. These measurements can be represented as the number pairs (t_1, P_1) and (t_2, P_2) plotted on pressure-time axes as two points. As everyone intuitively understands, there is only one straight line that can be drawn through these two points. The reason for this is that the equation of a straight line is characterized by two free parameters, its slope a and its intercept b. The values of a and b are determined by evaluating the equation for the line at the two points. This gives

$$P_1 = at_1 + b \tag{8.15}$$

and

$$P_2 = at_2 + b. \tag{8.16}$$

We thus have two simultaneous equations from which the two parameters a and b can be uniquely determined. The same argument applies to higher-order polynomials. Thus, for example, a quadratic function of pressure

$$P(t) = a_0 + a_1 t + a_2 t^2 \tag{8.17}$$

can be made to pass exactly through three data points because these points produce three simultaneous equations that can be solved to yield the values of the three parameters a_0, a_1, and a_2, and so on.

This procedure can be applied not just to polynomials, but to any function containing adjustable parameters. In particular, it can be applied to *sine waves*. The equation for a sine wave,

$$P(t) = A \sin(2\pi f t + \phi), \tag{8.18}$$

contains the three free parameters amplitude (A), frequency (f), and phase (ϕ). By the above reasoning, we should be able to fit a sine wave through three data points, two distinct sine waves through six points, and so on. However, sine waves are periodic, so if we find a sine wave that fits three data points, we can also achieve a fit by using higher harmonics of the same sine wave (Fig. 8.7). Therefore, before deciding how many sine waves are needed to make a curve that passes precisely through n points, some restrictions must be imposed on the frequencies that the sine waves can have.

In order to describe a set of data points as a sum of sine waves, we need to include a sine wave with a frequency of zero because this corresponds to a horizontal line. The level of this line can be adjusted to account for the mean value of the data. At the other end of the frequency scale, we need a sine wave that can change direction rapidly enough to account for point-by-point changes in the data values. On the other hand, having a sine wave change direction more than once between adjacent points is wasteful. The highest frequency sine wave needed to describe the data is thus one that oscillates through one

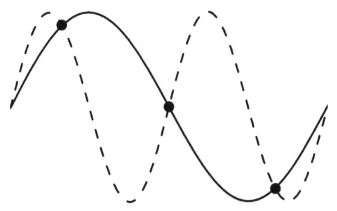

Figure 8.7 Fitting sine waves to data. Three data points (closed circles) can be intersected by a sine wave of lowest (fundamental) frequency (solid line), or by any of its higher harmonics (e.g. dashed line).

complete cycle with every two data points. This highest frequency is half that of the data itself. In order to complete the composite curve, sine waves of intermediate frequency must be added until the total number of free parameters equals the number of data points. By convention, these intermediate frequencies are chosen to be equally spaced between the two frequency extremes just identified.

Having fixed the frequencies of the sine waves, only their amplitudes and phases remain as free parameters. However, the zero-frequency sine wave has only one free parameter, its amplitude, because the phase of a horizontal line is meaningless. Also, the location of $t = 0$ along the time axis is arbitrary, but once it has been decided it can be viewed as determining the phase of the highest-frequency sine wave. The phases of the remaining sine waves are then defined relative to this reference phase.

As an example, consider the problem of constructing a composite sine wave function to pass through four data points. This obviously requires four free parameters. One parameter comes from the amplitude of the zero-frequency wave, and a second comes from the amplitude of the highest frequency wave. The remaining two parameters are provided by the amplitude and phase of a third sine wave having a frequency halfway between zero and half the data sampling rate. Figure 8.8 shows an example set of four data points together with the three individual sine waves and their sum that passes precisely through each point. For eight data points, three sine waves of intermediate frequency are required to bring the number of free parameters to eight. This process extends to any number of data points equal to an integer power of two.

The amplitudes and phases of the three sine wave components shown in Fig. 8.8 constitute the *discrete Fourier transform* of the four data points. Finding the values of these amplitudes and phases constitutes an exercise in solving simultaneous equations, the key step of which involves inverting a matrix of rank equal to the number of parameters. Matrix inversion is, in general, a computationally expensive process, but in

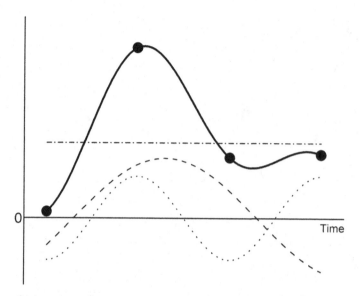

Figure 8.8 The discrete Fourier transform. Four equally spaced data points (circles) have a discrete Fourier transform consisting of three sine waves (broken lines), whose sum (solid line) passes exactly through all four points.

the case of the discrete Fourier transform the matrix to be inverted has a considerable degree of redundancy. This redundancy is exploited in the FFT algorithm, allowing the matrix to be inverted much more rapidly than would generally be the case. This is what makes the FFT "fast." The FFT algorithm was discovered in the 1960s [137] and revolutionized digital signal processing. Inverting a general matrix of rank n (i.e. n rows and n columns) takes of the order of n^2 arithmetic calculations, while calculating an FFT of the same rank takes of the order of $n\log_2(n)$ calculations. When n is large, the difference in computation time can be enormous (e.g. a factor of about 100 for 1000 data points).

The idea of fitting sums of sine waves to a finite number of discrete data points can be extended to the situation where the data sampling frequency tends to infinity and the individual points coalesce into a continuous curve. What were sets of discrete amplitudes and phases now become continuous functions of frequency. This gives rise to the *continuous Fourier transform* of the function $P(t)$, written as

$$P(f) = \int_{-\infty}^{\infty} e^{-i2\pi ft} P(t)\,dt$$

$$= \int_{-\infty}^{\infty} [\cos(2\pi ft) - i\,\sin(2\pi ft)]P(t)\,dt \tag{8.19}$$

where i is the positive square root of -1. $P(f)$ is a complex function of frequency with real and imaginary parts $R(f)$ and $X(f)$, respectively. Equation 8.19 can be inverted to

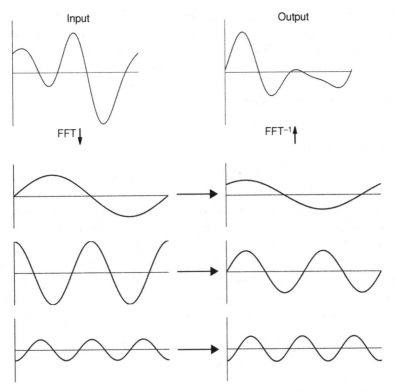

Figure 8.9 The action of a linear dynamic system on a general input signal. The signal shown in the top left panel is broken down into its component sine waves (lower left panels), each of which is then transformed by the system (lower right panels) and summed to produce the output (upper right panel).

give a formula for the *inverse Fourier transform* that provides $P(t)$ from $P(f)$ thus:

$$P(t) = \int_{-\infty}^{\infty} e^{i2\pi ft} P(f)\, df$$

$$= \int_{-\infty}^{\infty} [\cos(2\pi ft) + i\sin(2\pi ft)] P(f)\, df$$

$$= \int_{-\infty}^{\infty} [\cos(2\pi ft) + i\sin(2\pi ft)][R(f) + iX(f)]\, df. \qquad (8.20)$$

The reason for the great power of the Fourier transform in the analysis of linear systems is illustrated in Fig. 8.9. Using the forward FFT, an input can be decomposed into its constituent sine waves. Each of these sine waves can then be adjusted in amplitude and phase according to the frequency response of the system. Finally, the inverse FFT can be applied to the adjusted sine waves, and the result is the system output. Conversely, and potentially even more useful, this process can be reversed. That is, the FFT can be used to decompose a measured output into its individual sinusoidal components on which the effects of the frequency response can be "undone." The inverse FFT of the result then provides an estimate of the input. This procedure is known as *deconvolution*, and can be used, for example, to compensate for the effects of an imperfect frequency response in laboratory instrumentation [18].

8.2.2   The power spectrum

The fact that any signal can be decomposed into its constituent sine waves by the Fourier transform means that a signal can be characterized in terms of its *frequency content*. This refers to the frequencies that the signal contains and the relative contributions each frequency makes to the total signal. Frequency content is conventionally represented as a *power spectrum*, obtained by plotting the square of the amplitude of the Fourier transform against frequency. (The term "power spectrum" is derived from the fact that the power available to do work in a sinusoidally oscillating voltage is proportional to the square of its amplitude.) The square of the amplitude of the Fourier transform, being a complex quantity, is equal to the product of the transform with its complex conjugate, thus

$$|P(f)|^2 = [R(f) + iX(f)][R(f) - iX(f)]$$
$$= R^2(f) + X^2(f) \tag{8.21}$$

where R and X are the real and imaginary parts, respectively.

Figure 8.10 shows a signal in the time domain together with its frequency domain representation as a power spectrum. The time-domain representation looks like noise, but the power spectrum clearly shows the presence of two enhanced frequency components. This demonstrates that the time-domain and frequency-domain representations can bring out very different aspects of a signal even though the two representations are equivalent from the perspective of the amount of information each contains.

8.2.3   The convolution theorem for Fourier transforms

The power of the Fourier transform for analyzing linear dynamic systems derives in large part from the *convolution theorem*. This theorem states that when two functions of time are convolved in the time domain (as in Eq. 8.14), their respective Fourier transforms are multiplied in the frequency domain. This is shown by taking the Fourier transform of Eq. 8.14 using Eq. 8.19 thus:

$$F\left\{ \int_{-\infty}^{\infty} P(u)h(t-u)\,du \right\} = \int_{-\infty}^{\infty} \left[\int_{-\infty}^{\infty} P(u)h(t-u)\,du \right] e^{-i2\pi ft}\,dt$$

$$= \int_{-\infty}^{\infty} \int_{-\infty}^{\infty} P(u)h(t-u)e^{-i2\pi ft}\,dt\,du \tag{8.22}$$

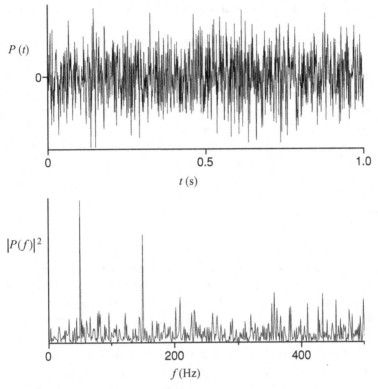

Figure 8.10 A time-domain signal (top panel) was generated by adding random noise to the sum of a sine wave at 50 Hz and another sine wave at 150 Hz. The presence of these two sine waves is difficult to discern in the time-domain representation, but is apparent as two spikes at the appropriate locations in the signal's frequency-domain representation as the power spectrum (bottom panel).

where F{ } denotes the operation of taking the Fourier transform of the bracketed quantity, and the second line of Eq. 8.22 is obtained by changing the order of integration. Now, replace $t - u$ with another variable, q, noting that $dt = dq$. Substituting q into Eq. 8.22 gives

$$
\int_{-\infty}^{\infty} \int_{-\infty}^{\infty} P(u)h(q)e^{-i2\pi f(q+u)}\, dq\, du = \int_{-\infty}^{\infty} P(u)e^{-i2\pi fu}h(q)e^{-i2\pi fq}\, dq\, du
$$

$$
= \int_{-\infty}^{\infty} P(u)e^{-i2\pi fu}\, du \int_{-\infty}^{\infty} h(q)e^{-i2\pi fq}\, dq
$$

$$
= P(f)H(f) \tag{8.23}
$$

where $P(f)$ and $H(f)$ are the Fourier transforms of $P(t)$ and $h(t)$, respectively. This proves the convolution theorem for Fourier transforms. When $h(t)$ is an impulse response function, its transform, $H(f)$, is called a *transfer function*. The transfer function of a linear dynamic system thus encapsulates all the information there is to know about the system.

Linear dynamic systems convert inputs to outputs in the time domain by the process of convolution, which is somewhat complicated to both compute and disentangle. What the above theorem shows is that in the frequency domain this same process reduces to simple multiplication. This provides an easy way to calculate what the output of a system will be from any input as illustrated in Fig. 8.9. That is, take the FFT of the input, multiply it by the FFT of the impulse response function, and then take the inverse FFT of the result. This calculation can also be performed in reverse to determine what the input is from a measured output (i.e. deconvolution).

8.3   Impedance

Impedance is the term given to a special kind of transfer function. What makes it special is the nature of the inputs and outputs involved. For example, when the input is *electric current* and the output is *voltage*, the transfer function connecting them is called *electrical impedance*. When input is *velocity* and output is *force*, the transfer function is called *mechanical impedance*. Current and velocity are both measures of how some quantity moves through a system. Voltage and force are corresponding measures of the extent to which such movement is hindered, or impeded. Impedance is thus a reflection of how *difficult* it is to get the quantity of interest to move through the system. Conversely, the transfer function between force as the input and velocity as the output is a measure of how easy it is to move through the system, and is called *admittance*.

In the field of lung mechanics, the relevant mechanical impedance is that between flow and pressure. Thus, if a flow signal $\dot{V}(t)$ is applied to the tracheal opening and the corresponding pressure signal $P(t)$ is measured also at the tracheal opening, the two signals are related by

$$P(f) = Z(f)\dot{V}(f) \qquad (8.24)$$

where $P(f)$ and $\dot{V}(f)$ are the Fourier transforms of $P(t)$ and $\dot{V}(t)$, respectively. $Z(f)$ in this case is termed *input impedance* because it relates two quantities measured at the entrance to the lung. By contrast, if $\dot{V}(t)$ had been applied to the surface of the thorax, perhaps by a pressurized vest, then $Z(f)$ would be called *transfer impedance* because it would relate two signals measured at different physical sites that span the respiratory system.

Equation 8.24 suggests that the determination of impedance should be straightforward: apply a suitable $\dot{V}(t)$ to the lungs, measure the resulting $P(t)$, take the FFT of both signals, and then perform the calculation

$$Z(f) = \frac{P(f)}{\dot{V}(f)}. \qquad (8.25)$$

In essence, this is precisely what is done, although the simplicity of Eq. 8.25 belies a number of practical challenges to which we now turn.

8.3.1   The forced oscillation technique

To determine $Z(f)$ using Eq. 8.25, a suitable $\dot{V}(t)$ must be applied to the lungs. One possibility is to make $\dot{V}(t)$ a sine wave at a particular frequency, in which case the division in Eq. 8.25 is performed only at that frequency. This can then be repeated at a number of different frequencies in turn. However, such a procedure is time-consuming, which limits temporal resolution and the ability to track transient mechanical responses in the lung. Fortunately, the principle of superposition and the FFT together make it possible to determine multiple frequency components of $Z(f)$ simultaneously. The way this is done is to make $\dot{V}(t)$ a *broad-band signal*, one that contains many different frequencies at the same time. So long as the lung behaves like a linear dynamic system, the resulting $P(t)$ signal will contain components at the same frequencies. Applying the FFT to $\dot{V}(t)$ and $P(t)$ produces transforms $\dot{V}(f)$ and $P(f)$ that can then be subjected to Eq. 8.25 at each of the frequencies they contain.

The *forced oscillation technique* is a general term for any approach that applies a broad-band $\dot{V}(t)$ signal to the lung for the purposes of measuring $Z(f)$. The forced oscillation technique has been in existence for more than 50 years [138] and is now the subject of a huge literature [139, 140]. Figure 8.11A depicts a typical experimental setup for applying the forced oscillation technique in a conscious subject [139, 141]. The oscillations in $\dot{V}(t)$ are produced with a loudspeaker and are measured with a pneumotachograph. $P(t)$ at the mouth is measured with a pressure transducer communicating with the flow stream through a lateral port. $\dot{V}(t)$ and $P(t)$ are low-pass filtered to avoid aliasing, followed by digitization at twice the filter cutoff frequency.

Flow oscillations can be applied to an *apneic* subject or over the top of *spontaneous breathing*. In the latter case, the frequency content of the applied oscillations must not overlap the *spectral content* of the breathing pattern. Most of the power in $\dot{V}(t)$ during spontaneous breathing resides at the fundamental frequency of breathing, which is typically below 0.5 Hz. The spontaneous breathing waveform is not a perfect sine wave, however, so there are often significant higher harmonics present at frequencies several times that of the fundamental. Forced oscillations in $\dot{V}(t)$ applied over spontaneous breathing thus typically begin at about 4 Hz.

The *wavetube* [142] provides an alternative method of applying the forced oscillation technique. Here, the oscillatory $\dot{V}(t)$ signal is directed into the subject's mouth via a length of rigid circular tubing. Lateral pressure is measured at two points along the length of the tube. A formula for the impedance of the tube between two pressure measurement sites is used to calculate the impedance of the mechanical load presented to the end of the tube [142]. The wavetube can be considered a complex pneumotachograph because it exploits not only differences in amplitude between two measured pressures, as does a conventional pneumotachograph, but also differences in phase.

Another important practical consideration for the measurement of input impedance is that the compliant portions of the upper airways (cheeks and accessible pharyngeal regions) must be supported. This forces the applied oscillations in $\dot{V}(t)$ to travel down into the lungs rather than being allowed to inflate these proximal structures. The conventional method of support is for the subject to press their hands against their cheeks. This does

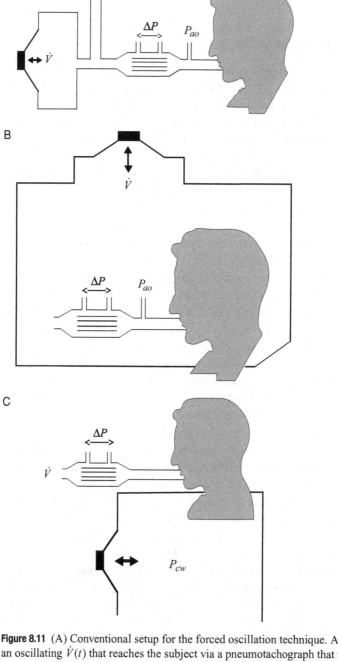

Figure 8.11 (A) Conventional setup for the forced oscillation technique. A loudspeaker generates an oscillating $\dot{V}(t)$ that reaches the subject via a pneumotachograph that measures a resistive pressure drop (ΔP) proportional to flow. Airway opening pressure (P_{ao}) is measured via a lateral tap. A high-impedance side tube between the loudspeaker and the pneumotachograph allows the subject to breathe while the forced oscillation technique is being applied. (B) The head generator method for applying the forced oscillation technique. A loudspeaker drives $\dot{V}(t)$ into a chamber sealed over the subject's head. The resulting pressure oscillations inside the mouth are nearly equal to those outside the head, preventing movement of the compliant cheeks and pharynx. (C) Setup for measuring transfer impedance. The subject sits inside a head-out body plethysmograph within which pressure oscillations around the chest wall (P_{cw}) are generated by a powerful loudspeaker. The resulting $\dot{V}(t)$ at the mouth is measured with a pneumotachograph.

not completely solve the problem of $\dot{V}(t)$ being shunted into the upper airways, however. An alternative, although more cumbersome, method is the so-called *head generator* [143]. Here, a rigid canopy is placed over the subject's head and sealed around the shoulders (Fig. 8.11B). A loudspeaker projects oscillations in $\dot{V}(t)$ into the canopy so that they travel into the subject's open mouth, but the pressures remain approximately equal either side of the upper airway tissues, which prevents shunting.

Figure 8.11C shows a setup for measuring transfer impedance [144]. Here, pressure oscillations are applied to and measured at the body surface while flow is measured at the airway opening. Note that it is $P(t)$ that is being controlled in this application, while $\dot{V}(t)$ is measured as the consequence. Nevertheless, we still formally consider $\dot{V}(t)$ to be the system input. Compared to input impedance, transfer impedance is relatively unaffected by central airway shunting and regional ventilation inhomogeneities [18].

The forced oscillation technique has also been applied to individual segments of the lung by forcing flow through the instrument channel of a clinical bronchoscope [145]. In dogs, tiny flow oscillations have been applied directly to sub-pleural alveoli through an alveoli capsule, enabling the impedance of a region roughly the size of a single acinus to be measured [74, 146]. It is not even necessary to use an external source of energy to generate oscillations in flow and pressure. For example, the energy of the heartbeat can be used to assess what has been termed the *output impedance* of the lung [147].

Finally, although Fig. 8.11 shows a loudspeaker as the device for actually generating oscillations in $\dot{V}(t)$, there are other possibilities. Some implementations of the forced oscillation technique use a variable-frequency piston pump [148, 149]. In fact, anything that causes flow to be other than a single sine wave is, in principle, suitable as a flow perturbation device for the forced oscillation technique. This includes ventilators with non-sinusoidal flow patterns [135, 150] and valves that suddenly interrupt flow either completely [132] or partially [151]. Indeed, it can be shown that the forced oscillation and flow interrupter techniques provide comparable information about lung mechanics [152].

8.3.2   A word about complex numbers

Dealing with impedance mathematically requires the use of complex numbers. At any given frequency, impedance itself is a complex number having both a real part and an imaginary part. This terminology is invariably a source of confusion when first encountered because it seems to suggest that part of the impedance (the imaginary part) is less relevant to the world we live in than the other part (the real part). This is not the case at all. The use of complex numbers is merely a mathematical trick that allows us to keep track of sine waves and cosine waves independently of each other. Equation 8.5 shows that when a sine wave goes into a linear dynamic system, both a sine wave and a cosine wave come out. In other words, part of the output (the sine wave part) is *in phase* with the input, while the remainder (the cosine wave part) is 90 degrees (or $\pi/2$ radians) *out of phase* with the input. The algebra of complex numbers merely provides a convenient formalism for carrying out this transformation and for keeping track of the in-phase and out-of-phase components.

8.3.3   Signal processing

The information about lung mechanics provided by the forced oscillation technique depends greatly on the frequency content of the $\dot{V}(t)$ that is applied to the lungs. Once this frequency range has been decided upon, however, there are various ways in which a broad-band $\dot{V}(t)$ signal can be realized. A classic method is to generate pseudo-random noise [153]. Another approach that has been used in a clinical device is to apply a train of square pulses in flow to the mouth [154]. The optimal approach from an engineering perspective, however, is to decide *a priori* exactly what frequency content the $\dot{V}(t)$ signal should have, and then generate precisely that signal. This gives rise to a so-called *composite signal* consisting of a sum of sine waves having frequencies chosen specifically to span the range of interest, with amplitudes chosen to give good signal-to-noise at each frequency. The phases of the sine waves can be chosen randomly, although a sensible scheme is to select them so as to minimize the peak–peak excursions in $\dot{V}(t)$ because the lower the amplitude of the perturbation, the more likely the lung is to behave like a linear system [155]. A widely used approach for spanning the frequency range of interest is to include all the harmonics of a fundamental frequency in the composite signal [156]. However, better signal-to-noise is generally obtained in the estimated $Z(f)$ if the frequencies are not harmonics of each other. The reasons for this will be covered in Chapter 11.

A particularly ingenious approach to generating composite flow signals for the forced oscillation technique is to adjust the amplitudes and phases of the component sine waves so that they comprise a waveform suitable for mechanically ventilating a subject. So long as the signal retains the necessary spectral content, it can still be used for the purposes of identifying $Z(f)$ [150]. An example of such an *optimal ventilator waveform* is shown in Fig. 8.12. The volume signal of the optimal ventilator waveform looks a bit shaky due to the higher frequency components it contains, but it still serves to deliver a functional tidal volume to a subject at a breathing frequency that is perfectly adequate for maintaining gas exchange.

Once $\dot{V}(t)$ and $P(t)$ have been collected, Eq. 8.25 would seem to suggest that calculating $Z(f)$ is then simply a matter of submitting them to the FFT algorithm to produce $\dot{V}(f)$ and $P(f)$, respectively, and then dividing $P(f)$ by $\dot{V}(f)$. In fact, there are some subtleties to this process that make it nontrivial. First, Eq. 8.25 can only be applied at frequencies where $\dot{V}(t)$ has significant power. If $\dot{V}(t)$ has been tailored to have a desired spectral content, then these frequencies are known. If not, the power spectrum of $\dot{V}(t)$ will first have to be determined so that the frequencies at which there is significant power can be identified.

Another important practical aspect of calculating $Z(f)$ concerns dealing with the inevitable noise that creeps into all measurements of $\dot{V}(t)$ and $P(t)$. The best way to reduce the effects of noise in any signal is to average multiple independent measurements of the signal, provided that the noise is random with respect to the signal. With the

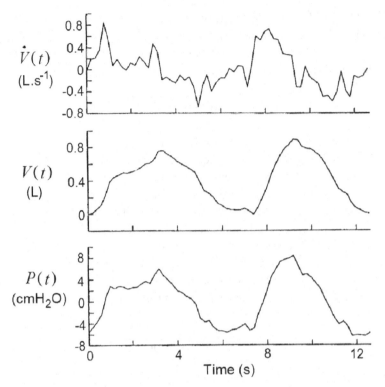

Figure 8.12 An example of an optimal ventilator waveform $\dot{V}(t)$ signal containing frequencies from 0.1 to 2 Hz (top). Its time-integral, $V(t)$, has the characteristics necessary for conventional ventilation of a human subject (middle), and produces a $P(t)$ tracing at the mouth similar to that normally seen with mechanical ventilation (bottom) (reproduced with permission from [150]).

forced oscillation technique, averaging is typically achieved with a perturbation signal that contains more than a single period (repeat length) of the flow waveform. The measurements of $\dot{V}(t)$ and $P(t)$ are divided into a series of overlapping windows, where the length of each window equals the period (T) of $\dot{V}(t)$ and the length of the overlap between adjacent windows is W [153, 157]. If $\dot{V}(t)$ contains N periods, then the number of overlapping windows is $(NT/W) - 1$. The FFTs of each windowed segment of $\dot{V}(t)$ and $P(t)$ are then used to calculate the average cross-power and auto-power spectral densities, given respectively by

$$C_{P,\dot{V}}(f) = \frac{W}{NT - W} \sum_{i=1}^{\frac{NT}{W}-1} P_i(f)\dot{V}_i^*(f) \qquad (8.26)$$

and

$$C_{\dot{V},\dot{V}}(f) = \frac{W}{NT - 1} \sum_{i=1}^{\frac{NT}{W}-1} \dot{V}_i(f)\dot{V}_i^*(f) \qquad (8.27)$$

where $P_i(f)$ and $\dot{V}_i(f)$ are the Fourier transforms of $P(t)$ and $\dot{V}(t)$, respectively, within the ith window, and $*$ denotes the complex conjugate. The greater the number of windows, the greater the reduction in noise achieved in the averaged power spectral densities. The densities are then divided to give the final estimate of impedance as

$$Z(f) = \frac{C_{P,\dot{V}}(f)}{C_{\dot{V},\dot{V}}(f)}. \tag{8.28}$$

The division in Eq. 8.28 is only performed, of course, at those frequencies at which $\dot{V}(t)$ has significant power (i.e. the frequencies of the sinusoidal components it contains). Calculating $Z(f)$ at other frequencies than these will result in division by zero, or something close to zero, which produces very large and erroneous results.

Equation 8.28 thus provides an average $Z(f)$ for all the data windows. This is not the same as the $Z(f)$ that would be obtained if the data from only a single window were used in Eq. 8.28. The reason is that ubiquitous noise and the inevitable nonlinearities that characterize any real-world system mean that $P(t)$ and $\dot{V}(t)$ are only imperfectly linked by a unique $Z(f)$. The extent of this imperfection is reflected in the variability of $Z(f)$ between different data windows and is quantified by the *coherence*, defined as

$$\Phi(f) = \frac{C^2_{P,\dot{V}}(f)}{C_{P,P}(f)C_{\dot{V},\dot{V}}(f)} \tag{8.29}$$

where $C_{P,P}(f)$ is the power spectral density of $P(t)$ defined in analogy with Eq. 8.27. $\Phi(f)$ is the frequency-domain analog of the CD (Eq. 3.31) used to assess goodness-of-fit for models in the time domain. When $P(t)$ and $\dot{V}(t)$ are linked precisely and completely over any time interval by a unique $Z(f)$ then $\Phi(f)$ has a value of 1.0 for all values of f. However, when noise and/or nonlinearities make extraneous contributions to $P(t)$, there will be at least some values of f at which $P(f)$ cannot be accounted for entirely by $Z(f)$ acting on $\dot{V}(f)$. At these f, the value of $\Phi(f)$ will be less than 1.0. If $P(t)$ and $\dot{V}(t)$ bear no relationship to each other at all then $\Phi(f)$ will be zero.

$\Phi(f)$ is particularly useful in lung mechanics applications for detecting when cardiogenic oscillations or spontaneous breathing interfere with the estimation of $Z(f)$. Both these events make a contribution to $P(t)$ that is independent of the $\dot{V}(t)$ signal applied during the forced oscillation technique, and therefore cause $\Phi(f)$ to fall. It then becomes a matter of defining a threshold value for $\Phi(f)$ below which $Z(f)$ is considered to be unacceptably corrupted. This threshold is somewhat arbitrary, although a value of 0.95 is typical for the measurement of lung impedance [158, 159].

Problems

8.1   A passive expiration can be viewed as the volume response of the lungs (or of the respiratory system, in the case of an intact chest wall) to a step change in pressure at the mouth. How will lung volume respond to a positive step change in pressure? What will the volume response be to a delta function in pressure?

8.2 The Fourier transform is presented in Section 8.2.1 in terms of fitting sums of sine waves to a collection of data points. Examine the formula for the Laplace transform and see if you can make a similar argument in terms of sums of decaying exponentials.

8.3 If a sine wave oscillating at 100 Hz is sampled at 120 Hz, what frequency does the sampled signal appear to have? What is the apparent frequency if the sampling rate is 80 Hz?

8.4 Using a pencil and paper, try to figure out what you get when you convolve a square pulse with itself. What do you get when you convolve the result with itself? Do you notice anything in particular about the result of repeated convolution?

8.5 Using the step response in pressure depicted in Fig. 8.5, sketch the pressure profile expected during square-wave positive pressure ventilation (i.e. when a mechanical ventilator provides a constant pressure during inspiration but allows expiration to be purely passive).

8.6 What does the impulse response corresponding to the step response depicted in Fig. 8.5 look like?

8.7 Explain why integrating a signal with respect to time is equivalent to convolving the signal with the Heaviside step function of time. What function is a signal convolved with to achieve differentiation?

9   Inverse models of lung impedance

Lung impedance is an empirical characterization of how the organ functions mechanically. Even though impedance contains everything there is to know about the system from a mechanical point of view (provided, of course, it is a linear dynamic system), the usefulness of such information is often not immediately obvious. What we really need to know is how the information embodied in the mechanical impedance of the lung relates to its internal structure. This is a more general statement of the problem we faced in Chapter 3 in contemplating the utility of measurements of lung resistance and elastance. In the case of impedance, mathematical models again provide the way forward.

9.1   Equations of motion in the frequency domain

In this chapter we will examine the impedances of the one and two degree-of-freedom models developed in Chapters 3 and 7. Obviously, we are not going to submit the nonlinear models in Chapter 5 to this analysis because, being nonlinear, the impedance concept is not relevant. First, however, an important mathematical preliminary must be established. To calculate the impedance of a linear lumped-parameter model, it will be necessary to determine the Fourier transform of the model's equation of motion. This equation, as we have seen in Chapter 8, is a linear differential equation that is expressed explicitly in $P(t)$ in the general form

$$P(t) = \sum_{n=0}^{N} a_n \frac{d^n V}{dt^n} + \sum_{n=1}^{N-1} b_n \frac{d^n P}{dt^n} \tag{9.1}$$

for a model of N degrees of freedom, where the a_n and b_n are constants composed of algebraic combinations of the parameters that represent the constitutive properties of the model. Calculating the Fourier transform of Eq. 9.1 thus involves calculating the transforms of time-derivatives of $P(t)$ and $V(t)$. This requires the following Fourier identity:

$$F\left\{\frac{dV(t)}{dt}\right\} = \int_{-\infty}^{\infty} e^{-i\omega t}\frac{dV(t)}{dt}\,dt$$

$$= \left[e^{-i\omega t}V(t)\right]_{-\infty}^{\infty} - \int_{-\infty}^{\infty} -i\omega e^{-i\omega t}V(t)\,dt$$

$$= i\omega \int_{-\infty}^{\infty} e^{-i\omega t}V(t)\,dt$$

$$= i\omega F\{V(t)\} \qquad (9.2)$$

where we have used the standard notation for *angular frequency*, $\omega = 2\pi f$. The second line of Eq. 9.2 was obtained from the first line using integration by parts. The third line was obtained from the second assuming that the function being transformed, $V(t)$ in this case, is "well behaved," meaning that it has the decency to become zero as time tends to either plus or minus infinity. Equation 9.2 shows that the Fourier transform of the time-derivative of a function is simply the transform of the function itself multiplied by $i\omega$. Transforms of higher-order derivatives are obtained by multiplication by the factor $i\omega$ raised to the requisite power. Integration, being the opposite of differentiation, is dealt with by multiplying the transform of $V(t)$ by the inverse of the above factor, namely $-i/\omega$. The Fourier transform of Eq. 9.1 is thus

$$P(\omega) = \left[\sum_{n=0}^{N}(i\omega)^n\,a_n\right]V(f) + \left[\sum_{n=1}^{N-1}(i\omega)^n\,b_n\right]P(\omega). \qquad (9.3)$$

The impedance of the model is obtained by rearranging this equation to give

$$Z(\omega) = \frac{P(\omega)}{i\omega V(\omega)}$$

$$= \frac{\displaystyle\sum_{n=0}^{N}(i\omega)^{n-1}\,a_n}{1 - \displaystyle\sum_{n=1}^{N-1}(i\omega)^n\,b_n}. \qquad (9.4)$$

9.2 Impedance of the single-compartment model

The equation of motion of the single-compartment linear model with resistance R and elastance E (Section 3.1.2) is

$$P(t) = R\dot{V}(t) + EV(t) + P_0. \qquad (9.5)$$

Remembering that $V(t)$ is the time-integral of $\dot{V}(t)$ and invoking the above Fourier identity concerning the transform of an integral, we find the Fourier transform of

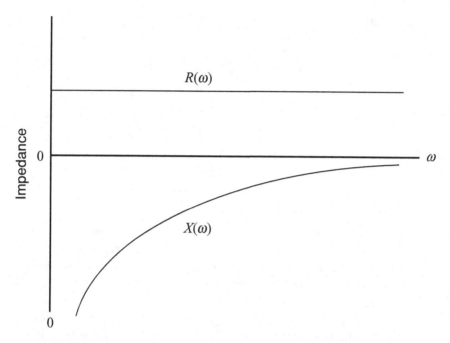

Figure 9.1 Resistance (R) and reactance (X) of the single-compartment linear model of the lung as a function of angular frequency (ω).

Eq. 9.5 to be

$$P(\omega) = R\dot{V}(\omega) - \frac{i}{\omega}E\dot{V}(\omega)$$

$$= \left[R - \frac{iE}{\omega}\right]\dot{V}(\omega). \qquad (9.6)$$

The input impedance of the single-compartment linear model is therefore

$$Z(\omega) = \frac{P(\omega)}{\dot{V}(\omega)}$$

$$= R(\omega) + iX(\omega)$$

$$= R - \frac{iE}{\omega}. \qquad (9.7)$$

The real part of this impedance, $R(\omega)$, is equal to R for all values of ω. In fact, $R(\omega)$ is called *resistance* for this reason. The imaginary part, $X(\omega)$, is called *reactance* and is a negative hyperbolic function of ω (Fig. 9.1).

9.2.1   Resonant frequency and inertance

When oscillation frequencies exceed those of normal breathing, the pressures required to accelerate structures in the lung begin to become important. The principal contributor to this phenomenon is the mass of the gas in the central airways, which can be viewed to a first approximation as a cylinder of gas occupying a conduit of length l

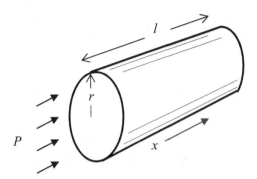

Figure 9.2 An airway of length l and radius r contains a mass of gas $\pi r^2 l \rho$. This mass is accelerated in the axial (x) direction by pressure P.

and radius r (Fig. 9.2). If the gas has density ρ then the mass of the gas is $\pi r^2 l \rho$. The force acting to accelerate this cylinder is the pressure (P) acting over the cross-sectional area of the conduit. The acceleration of the cylinder in the axial direction (dx/dt) is equal to the volume acceleration divided by the conduit cross-section. Substituting these components into Newton's law of motion gives

$$\pi r^2 P = \pi r^2 l \rho \left(\frac{1}{\pi r^2} \frac{d^2 V}{dt^2} \right). \tag{9.8}$$

In other words,

$$P = \frac{l\rho}{\pi r^2} \ddot{V}$$
$$= I\ddot{V}. \tag{9.9}$$

The coefficient of $\ddot{V}$ in Eq. 9.9 is called *inertance*, and is given by

$$I = \frac{l\rho}{\pi r^2}. \tag{9.10}$$

I thus increases as the airways narrow. This might seem a bit counterintuitive given that narrower airways would contain a smaller mass of gas, but this is merely a consequence of the fact that Eq. 9.8 is expressed in terms of pressure and flow rather than force and velocity.

The inertive term can be added to the equation of the single-compartment model to yield a second-order differential equation of motion, thus:

$$P(t) = I\ddot{V}(t) + R\dot{V}(t) + EV(t) + P_0. \tag{9.11}$$

Taking the Fourier transform of Eq. 9.11 yields the input impedance

$$Z(\omega) = R + i \left[\omega I - \frac{E}{\omega} \right]. \tag{9.12}$$

I is usually rather small because air does not have much mass, so the inertive contribution to $Z(\omega)$ only becomes significant at high frequencies when accelerations are

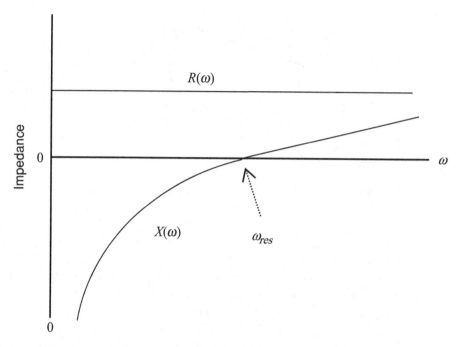

Figure 9.3 Resistance (R) and reactance (X) of the single-compartment linear model with inertance in the airway. ω_{res} is the resonant frequency.

large. Accordingly, at low ω the magnitude of E/ω in Eq. 9.12 is much larger than the magnitude of ωI, so $X(\omega)$ is negative. As ω increases, however, the term containing I becomes progressively larger while the magnitude of the term containing E decreases. Eventually the term containing I will dominate. $X(\omega)$ is zero when E/ω and ωI are equal. The value of ω at which this occurs is called the *resonant frequency*, ω_{res} (Fig. 9.3). From Eqs. 9.10 and 9.12 we see that

$$\omega_{res} = \sqrt{\frac{E}{I}}$$

$$= r\sqrt{\frac{\pi E}{l\rho}}. \tag{9.13}$$

A considerable amount of physiological information can be inferred about abnormalities in lung mechanical function from the way that input impedance varies with frequency between different patient groups. For example, Fig. 9.4 shows average real (R) and imaginary (X) parts of input impedance measured in isolated human lungs from healthy subjects, senile subjects who had some degree of airspace enlargement, and patients with emphysema [60]. Over the frequency range 2 to 32 Hz shown in this plot, R for both the healthy and senile subjects is virtually independent of frequency. This indicates that the lungs of these subjects behaved much like a single uniformly ventilated compartment, as represented in Fig. 9.3. Subjects with emphysema had markedly increased degrees of airway obstruction, as shown by a greatly increased R. Furthermore,

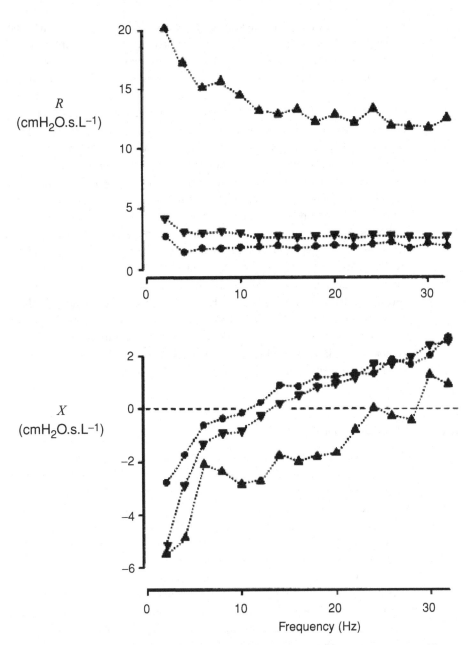

Figure 9.4 Real and imaginary parts (R and X, respectively) of input impedance measured in isolated human lungs. Normal lung = circles; senile lung = triangles pointing down; emphysematous lung = triangles pointing up (adapted with permission from [60]).

these subjects exhibit a marked negative dependence of R on f, which is the signature of heterogeneous mechanical behavior throughout the lungs, as explained in Chapter 7. X in the healthy subjects in Fig. 9.4 also corresponds with Fig. 9.3, and crosses the horizontal axis at the expected resonant frequency for healthy adult humans of about 8 Hz. The resonant frequency is increased slightly in senile lungs and markedly in emphysema, which Eq. 9.13 tells us to expect when airway radius (r) is reduced. This effect can be attributed to the progressive destruction of the lung parenchyma that begins in old age and is greatly exaggerated in emphysema. Specifically, when the parenchymal attachments that normally tether open the airways are broken, the airway wall is free to collapse inward, which leads to an increase in airway resistance. The impedances in Fig. 9.4 thus paint a consistent picture that links heterogeneous airway narrowing to increasing destruction of the lung parenchyma.

The single-compartment model with an airway inertance is thus capable of accounting for many, but not all, features in the input impedance spectrum of the respiratory system over the intermediate frequency range shown in Fig. 9.4. An important exception is the monotonic decrease in R with increasing f evident in the patient groups. This is not something that can be explained by the single-compartment model, whether it contains an airway inertance or not, because this model predicts that R should be constant with f. Explaining the negative frequency dependence of R requires models with more than one compartment. Before moving on to consider such models, however, we will examine one more application of the single-compartment model.

9.2.2   Regional lung impedance

So far we have been considering models of the entire lung, as is appropriate for situations in which input impedance is measured via the mouth or tracheal opening. In this case, the forced oscillations in flow are distributed to all regions of the lung in inverse proportion to their respective local impedances. It is possible, however, to apply flow oscillations to only selected regions. For example, if a bronchoscope is advanced into the airway tree to the point where its distal end meets an airway of matching diameter, the scope can be wedged into the airway so that it isolates the downstream segment of the lung. Forced oscillations in flow applied through the instrument channel of the scope can then be used to determine the input impedance of only the wedged segment [145].

Alternatively, it is possible to measure the impedance of an even smaller segment of the lung by applying tiny flow oscillations through an alveolar capsule [146]. The model that has been shown to describe this *alveolar input impedance* (Z_A) up to 200 Hz is shown in Fig. 9.5A. This model features a single elastic compartment representing the stiffness (E_A) of an alveolar region of the lung just under the pleural surface. The hole in the pleural surface is represented by a short conduit with resistance R_{hole} and inertance I_{hole}, while the connection of the compartment to the rest of the lung is represented by a conduit of resistance R_A. Modeling studies have shown that E_A has a value close to the stiffness expected of a single acinus, while R_A is close to the resistance expected of a terminal airway [160, 161].

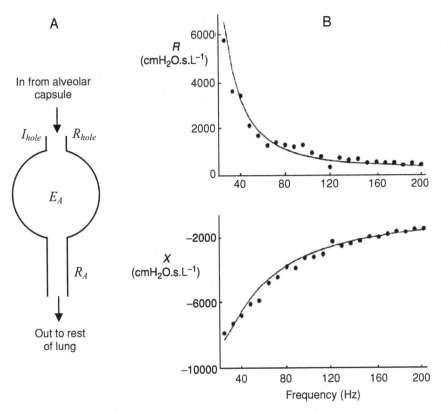

Figure 9.5 (A) Model of sub-pleural lung region connected to an alveolar capsule oscillator via a short conduit with resistance R_{hole} and inertance I_{hole}. E_A is the elastance of a region similar in size to a single acinus, while R_A is the resistance of the terminal airway serving it. (B) Example of Z_A up to 200 Hz in a dog and the fit provided by the model (reproduced with permission from [74]).

The impedance of the model in Fig. 9.5A is [146]

$$Z_A = \frac{E_A^2(R_{hole} + R_A) + \omega^2 R_{hole} R_A^2 + i\left[\omega^3 R_A^2 I_{hole} + \omega\left(I_{hole} E_A^2 - R_A^2 E_A\right)\right]}{E^2 + \omega^2 R_A^2}.$$

(9.14)

Equation 9.14 is based on the assumption that the lung, because of its relatively much larger volume, behaves like an infinite sink for the flow exiting the terminal airway. A useful check on the validity of Eq. 9.14 is to imagine that the terminal airway is blocked (i.e. R_A is infinite). In this case, the model becomes topologically equivalent to the single-compartment model of the whole lung. Accordingly, Eq. 9.14 adopts the same functional form as Z for the entire lung (Eq. 9.12) as R_A tends to infinity.

Figure 9.5B shows an example of Z_A up to 200 Hz measured in a dog, together with the fit provided by Eq. 9.14. The fitted curve captures the overall features of the data very well. These model fits thus give values for R_A and E_A that characterize the overall response of a tiny peripheral region of the lung during bronchoconstriction. An intriguing aspect of this response is that it varies greatly from one region to another, shown by simultaneous measurements from two capsule oscillators installed on the same lung [74, 162]. Despite this variability, however, which is both spatial and temporal in nature, the impedance of the entire lung changes in a predictable and reproducible way during challenge with smooth muscle agonists. This demonstrates that the mechanical properties of the lungs manifest in different ways at different levels of scale.

9.3   Impedance of multi-compartment models

The single-compartment model with an airway inertance considered above is described by a second-order differential equation (Eq. 9.11). This means that although this model considers the lung to have only a single alveolar compartment, it actually has two mechanical degrees of freedom. That is, to uniquely specify the state of the model at any instant, two independent quantities are required. These quantities are the volume of the compartment and the velocity of the air in the airway. In Chapter 7 we discussed three other plausible models of lung mechanics that have two degrees of freedom – the parallel model (Fig. 7.2), the series model (Fig. 7.3), and the viscoelastic model (Fig. 7.5). All are described by second-order differential equations.

9.3.1   The viscoelastic model

Of the three models considered in Chapter 7, the viscoelastic model has been shown to be the most appropriate for describing the normal lung over the range of frequency corresponding to breathing. The equation describing the viscoelastic model is, from Section 7.3,

$$R_t \dot{P}(t) + E_2 P(t) = R_{aw} R_t \ddot{V}(t) + (E_1 R_t + E_2 R_{aw} + E_2 R_t) \dot{V}(t) + E_1 E_2 V(t).$$
(9.15)

Expressing the Fourier transform of this equation in terms of $P(f)$ and $\dot{V}(t)$ gives

$$[i\omega R_t + E_2] P(f) = \left[i\omega R_{aw} R_t + (E_1 R_t + E_2 R_{aw} + E_2 R_t) - \frac{i E_1 E_2}{\omega} \right] \dot{V}(t),$$
(9.16)

from which we obtain

$$Z(\omega) = \frac{E_2^2 (R_{aw} + R_t) + \omega^2 R_{aw} R_t^2}{E_2^2 + \omega^2 R_t^2} - i \left[\frac{\omega^2 R_t^2 (E_1 + E_2) + E_1 E_2^2}{\omega E_2^2 + \omega^3 R_t^2} \right].$$
(9.17)

The real and imaginary parts in Eq. 9.17 are still referred to as resistance and reactance, respectively, because at any particular value of f there is an equivalent single-compartment model having the same values of resistance and reactance. Furthermore, in analogy with Eq. 9.7, we can calculate the apparent elastance of the viscoelastic model at any given frequency by multiplying reactance by $-\omega$. The resistance and elastance of

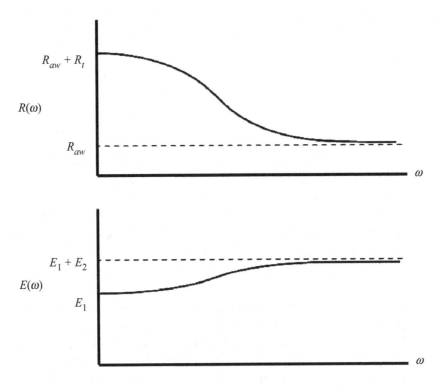

Figure 9.6 Resistance (top) and elastance (bottom) versus frequency for the viscoelastic model (Fig. 7.5).

the viscoelastic model are thus both functions of f, given by

$$R(\omega) = \frac{E_2^2\,(R_{aw}+R_t)+\omega^2 R_{aw}\,R_t^2}{E_2^2+\omega^2\,R_t^2} \tag{9.18}$$

and

$$E(\omega) = \frac{\omega^2\,R_t^2\,(E_1+E_2)+E_1 E_2^2}{E_2^2+\omega^2\,R_t^2}. \tag{9.19}$$

Inspection of Eq. 9.18 shows that $R(0)$ equals the sum of R_{aw} and R_t, while $R(\infty)$ is simply R_{aw} alone. Similarly, from Eq. 9.19 we see that $E(0) = E_1$ and $E(\infty) = E_1 + E_2$. Between these two frequency extremes, R and E decrease and increase, respectively, in a monotonic sigmoidal fashion as illustrated in Fig. 9.6. Consequently, we say that the resistance and elastance of this model are *frequency dependent*.

9.3.2 Effects of ventilation heterogeneity

The parallel (Fig. 7.2) and series (Fig. 7.3) two-compartment models yield equations for $R(\omega)$ and $E(\omega)$ that exhibit precisely the same kind of dependence on frequency as those embodied in Eqs. 9.18 and 9.19. We thus have two possible explanations for

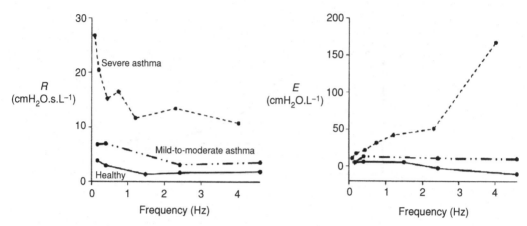

Figure 9.7 Resistance (left) and elastance (right) up to 5 Hz in a healthy subject, a subject with asthma of intermediate severity, and a severe asthmatic (reproduced from [149] and from [167] with kind permission of Springer Science and Business Media).

the decreasing real part of impedance seen in Fig. 9.4, namely viscoelasticity of the respiratory tissues and regional heterogeneity of ventilation. It is not possible to distinguish between these two possibilities from inspection of the frequency dependence of impedance alone. However, the fact that the frequency dependence in Fig. 9.4 is increased in the diseased subjects makes ventilation heterogeneity a more likely candidate in this case. Also, studies with anatomically based computational models of the lung strongly support the notion that ventilation heterogeneity is responsible for most frequency dependence of resistance observed over the intermediate frequency range shown in Fig. 9.4 [47, 145, 163]. It must be pointed out, however, that some studies of normal lungs have shown resistance to increase slightly with frequency over the range illustrated in Fig. 9.4 [164]. This is due to airway gas inertia, which can have several effects, including changing the distribution of flow between parallel pathways and altering flow velocity profiles in a frequency dependent manner.

Tissue viscoelasticity has a substantial influence on the frequency dependence of impedance, but these effects are concentrated below about 2 Hz because the dynamics of stress adaptation in tissue are slow. Measurements of impedance made in animals using the alveolar capsule technique [135, 165, 166] and in subjects breathing gases of different physical properties [73] have confirmed that these low-frequency variations in $R(\omega)$ and $E(\omega)$ are almost entirely due to the viscoelastic properties of the tissues in the healthy lung. In pathologies such as asthma, however, mechanical malfunctions of the lung are invariably manifest in a very heterogeneous manner. This can lead to greatly exaggerated variations of $R(\omega)$ and $E(\omega)$ with frequency.

Figure 9.7 shows examples of the low-frequency impedance patterns that can arise when the airways become narrowed [149]. In mild-to-moderate asthma, the entire $R(\omega)$ curve has been shifted upward relative to the healthy curve, suggesting that the resistance of the airway tree has been increased by a relatively uniform narrowing of all its individual airway segments. The corresponding $E(\omega)$ curve has similarly been shifted vertically,

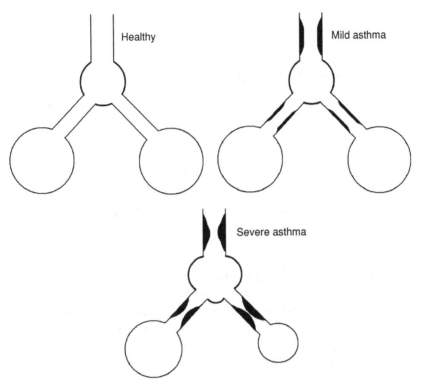

Figure 9.8 Compartmentalization of the lung in asthma. The healthy lung is represented as a pair of identical alveolar compartments with a stiff central airway compartment that does not become inflated when flow is forced into the alveolar compartments. In mild asthma, a modest degree of airway narrowing still allows the alveolar compartments to inflate relatively uniformly. In severe asthma, the imposed flow has to inflate the central airway compartment before it is able to pass into the alveolar compartments, which themselves inflate non-uniformly.

indicating a generalized increase in the stiffness of the tissues, but retains its downward trend at higher frequencies due to the increasing effects of airway inertance. By contrast, in severe asthma, not only is $R(\omega)$ shifted up to markedly higher values, the negative dependence on frequency is also greatly exaggerated, indicative of an increase in the heterogeneity of ventilation. This can be explained by a highly heterogeneous change in the resistances of individual airways, some remaining almost normal while others come close to closing completely. The corresponding $E(\omega)$ curve shows a progressive increase with frequency. This can be explained by a phenomenon known as *central airway shunting* [48, 149] in which the imposed oscillations in flow used to measure impedance become progressively less able to travel into the distal regions of the lung as frequency increases. In other words, the heterogeneities in this lung are not only parallel in nature like the model in Fig. 7.2, but also serial as in Fig. 7.3.

Modeling the impedance of the constricted lung thus requires a combination of both series and parallel compartments [168]. At the simplest level, this means a model of three compartments as shown in Fig. 9.8. The proximal compartment represents the

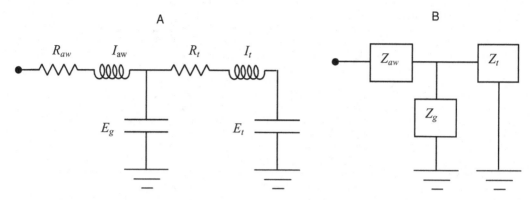

Figure 9.9 (A) Electrical circuit representation of the six-element model. (B) The T-network representation. Note that the capacitors are labeled as elastances, which are the inverses of the capacitances conventionally used in electric circuits.

distensibility of the airway tree, while the two distal compartments represent disparate alveolar regions. When the lung is normal and the airways are open (Fig. 9.8A), low-frequency oscillations in flow imposed at the airway opening are able to travel uniformly into the two alveolar compartments without causing measurable inflation of the much stiffer airway compartment. In mild asthma, the airways become slightly narrowed (Fig. 9.8B), which increases their overall resistance but still allows virtually all the imposed flow to reach the alveolar compartments. In severe asthma when the distal airways become markedly and heterogeneously narrowed, flow entering the airways has to first expand the proximal airway compartment in order to generate the pressures necessary to pass through to the two alveolar compartments. The extent of this central airway shunting becomes progressively more pronounced as frequency increases, causing the overall stiffness of the lung to look progressively more like that of the stiff central airways.

9.3.3   The six-element model

Most early investigations with the forced oscillation technique applied to the lung utilized an intermediate range of frequencies beginning slightly above the breathing frequency range (typically around 4 Hz) and ending at 30–60 Hz [153, 164]. This frequency range has the advantage of being above that of normal respiration, so subjects can breathe while measurements are being made without the impedance spectra becoming corrupted. Also, the resonant frequency predicted by Eq. 9.13 occurs within this frequency range in a normal adult human, allowing the parameters of the single-compartment model to be identified reliably [169]. Even from the very beginnings of the forced oscillation technique, however, it became popular to think of the lung in terms of a model of six elements based on those structures in the lung that were thought to be important for determining impedance [138]. This model is not easily represented as a mechanical structure of lumped elements, like the compartment models in Chapter 7. It is therefore usually represented as the electrical circuit shown in Fig. 9.9A.

The six-element model includes not only the airway resistance (R_{aw}), airway inertance (I_{aw}) and tissue elastance (E_t) of the single-compartment model, but also the resistance (R_t) and inertance (I_t) of the lung tissues. The model also accounts for the fact that not all the flow leaving the airways is translated into movement of the tissues; some of it is lost to compression of alveolar gas, which has elastance E_g.

The six-element model has four degrees of freedom, one each for the two inductors and the two capacitors in Fig. 9.9A. Therefore, its equation of motion is a fourth-order linear differential equation. In principle, one can derive this equation using the methods outlined in Section 7.2, and then take the Fourier transform of the equation to obtain impedance. However, as we have seen, the algebraic machinations required to perform this task are already bad enough for a model with only two degrees of freedom. As the number of degrees of freedom increases above two, the algebra rapidly becomes overwhelming.

Fortunately, there is a more efficient way to calculate the impedance of a lumped-parameter model, such as that in Fig. 9.9A, that does not involve its time-domain differential equation. This method is to add the component impedances of the model together directly in the frequency domain. To do this, the various model components must first be segmented into three distinct sub-impedances constituting a T-network as shown in Fig. 9.9B. The elements R_{aw} and I_{aw} comprise the impedance of the airways (Z_{aw}), R_t, I_t, and E_t comprise tissue impedance (Z_t), and E_g on its own comprises the impedance of the alveolar gas (Z_g). These three sub-impedances are all versions of Eq. 9.12, so we can write down their equations immediately as

$$Z_{aw}(\omega) = R_{aw} + i\omega I_{aw}, \tag{9.20}$$

$$Z_t(\omega) = R_t + i\left[\omega I_t - \frac{E_t}{\omega}\right], \tag{9.21}$$

and

$$Z_g(\omega) = \frac{-iE_g}{\omega}. \tag{9.22}$$

Now consider what happens when current enters the complete model (Fig. 9.9B) and passes through to ground. All of the current has to pass through Z_{aw}, but it is divided between Z_g and Z_t. This means that Z_g and Z_t act in parallel, but together they are in series with Z_{aw}. Impedances add together in series and parallel in exactly the same way as do resistances (Section 4.1.4), except that the rules of complex algebra apply because impedances are complex numbers that have both real (R) and imaginary (X) parts. The input impedance of the entire model is thus

$$
\begin{aligned}
Z_{in} &= Z_{aw} + \frac{Z_g Z_t}{Z_g + Z_t} \\
&= (R_{aw} + iX_{aw}) + \frac{(R_g + iX_g)(R_t + iX_t)}{(R_g + iX_g) + (R_t + iX_t)} \\
&= (R_{aw} + iX_{aw}) + \frac{(R_g R_t - X_g X_t) + i(R_g X_t + X_g R_t)}{(R_g + R_t) + i(X_g + X_t)}.
\end{aligned}
\tag{9.23}
$$

To express Eq. 9.23 as a sum of separate real and imaginary parts, all denominators must be made purely real. This is achieved by multiplying top and bottom by the complex conjugate of the denominator, thus

$$
\begin{aligned}
Z_{in} &= (R_{aw} + iX_{aw}) \\
&\quad + \frac{(R_g R_t - X_g X_t) + i(R_g X_t + X_g R_t)}{(R_g + R_t) + i(X_g + X_t)} \frac{[(R_g + R_t) - i(X_g + X_t)]}{[(R_g + R_t) - i(X_g + X_t)]} \\
&= \left[R_{aw} + \frac{R_t(R_g^2 + X_g^2) + R_g(R_t^2 + X_t^2)}{(R_g + R_t)^2 + (X_g + X_t)^2} \right] \\
&\quad + i\left[X_{aw} + \frac{X_t(R_g^2 + X_g^2) + X_g(R_t^2 + X_t^2)}{(R_g + R_t)^2 + (X_g + X_t)^2} \right].
\end{aligned} \tag{9.24}
$$

Equations 9.20 through 9.22 can now be substituted into this expression to yield the final equation for the impedance of the six-element model. The result is a rather complicated expression, and it would get even more complicated if it became necessary to divide each of the three sub-impedances Z_{aw}, Z_t, and Z_g in Fig. 9.9B into their own respective T-networks. Again, keeping track of the entire expression by hand rapidly becomes overwhelming. However, it is imminently feasible to calculate the impedances of complicated networks using recursive algorithms implemented on a computer. Indeed, this approach has been used to calculate the impedance of anatomically based numerical models of the lung that include every airway in the entire tree [47, 48, 170].

The six-element model has two inertive elements so it predicts that $Z(\omega)$ should exhibit two distinct resonances. We have already mentioned the first, which occurs around 8 Hz in a healthy human adult and is due to ringing (damped oscillations) of the mass of the gas in the conducting airways as it interacts with lung elastance. The second resonance is due to the mass of the lung tissue, and should occur at a much higher frequency. Such a resonance is observed in dogs as a peak in $R(\omega)$ at around 80 Hz [132, 163], but seems to be over-damped in humans [163].

9.3.4   Transfer impedance

The six-element model has also been applied to the interpretation of transfer impedance, which can be determined from the relationship between either flow and pressure at the airway opening and pressure at the body surface, or alternatively pressure at the airway opening and pressure at the body surface [159, 171]. Conventionally, a body plethysmograph has been used to make the body surface measurements in either case [159, 172] (Fig. 8.11C), although a more technically complicated but less invasive optical technique has been used recently [173, 174].

The transfer impedance, Z_{tr}, of the six-element model is obtained by noting that the flow, $\dot{V}$, which passes through the airway compartment becomes divided between the gas and tissue compartments. As the pressure drop across both compartments is the same, this gives

$$\dot{V}_t(\omega)Z_t(\omega) = \dot{V}_g(\omega)Z_g(\omega)$$
$$= \left[\dot{V}(\omega) - \dot{V}_t(\omega)\right]Z_g(\omega), \qquad (9.25)$$

from which we obtain

$$\dot{V}_t(\omega) = \frac{Z_g}{Z_t + Z_g}\dot{V}(\omega). \qquad (9.26)$$

Z_{tr} is the frequency-domain relationship between pressure at the airway opening (P) and $\dot{V}_t$. Therefore, from Eqs. 9.23 and 9.26,

$$
\begin{aligned}
Z_{tr}(\omega) &= \frac{P(\omega)}{\dot{V}_t(\omega)} \\
&= \frac{(Z_t + Z_g)P(\omega)}{Z_g\dot{V}(\omega)} \\
&= \frac{Z_t + Z_g}{Z_g}\left[Z_{aw} + \frac{Z_g Z_t}{Z_g + Z_t}\right] \\
&= Z_{aw} + Z_t + \frac{Z_{aw}Z_t}{Z_g}. \qquad (9.27)
\end{aligned}
$$

The above analysis shows that, according to the six-element model, it should be possible to predict Z_{tr} (Eq. 9.27) from Z_{in} (Eq. 9.23), and *vice versa*. In fact, the agreement between the two from 4 to 30 Hz is quite good (Fig. 9.10), but it is not perfect. Of course, perfection is not expected because the lung is a complex organ and a model of only six parameters is sure to miss something important. Regional heterogeneities of ventilation are an obvious candidate, and indeed have been invoked to account for the observed discrepancies [159, 175].

Even without the presence of heterogeneities, in order to obtain statistically reliable estimates of the parameters of the six-element model, it is necessary to fit the model to measurements of impedance made over a sufficiently wide range of frequencies. This frequency range should include data close to the second resonance [176], which is not far below 100 Hz, even when both input and transfer impedances are fit simultaneously [177]. One might naturally assume, therefore, that fitting the six-element model to data over an even greater frequency range would yield more accurate and robust parameter estimates. However, as frequency approaches 200 Hz, a new phenomenon begins to appear, because at these frequencies the wavelength of sound approaches the dimensions of the lung itself. This renders lumped-parameter models, and their ordinary differential equations of motion, inappropriate for describing lung mechanics. At high oscillation frequencies, it is therefore necessary to invoke continuous-parameter models that are described by partial differential equations in which both time and distance appear as independent variables. We have already seen an example of this in Chapter 6, where it was necessary to invoke the wave equation to account for the movement of elastic waves in the airway wall. In the case of high-frequency lung impedance, it is necessary to account for the propagation of acoustic waves within the airway lumen.

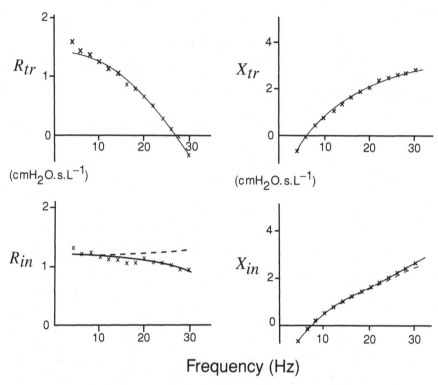

Figure 9.10 Transfer impedance (top: resistance R_{tr}, reactance X_{tr}) and input impedance (bottom: resistance R_{in}, reactance X_{in}) in a human subject. Symbols ($\times$) show experimental data points. Solid lines show fits of the six-element model. Dashed lines in bottom panels show the prediction of input impedance based on fits to transfer impedance (reproduced with permission from [159]).

9.4   Acoustic impedance

The forced oscillation technique can be applied to the lungs over whatever frequency range the equipment employed is capable of generating oscillations in flow. Flow oscillations can easily be generated far into the acoustic range by loudspeakers that are, of course, built expressly for this purpose. The impedance of the lungs has been well studied up to several hundred Hz, and has been shown to depend markedly on the physical properties of the gas in the airways [132, 178], as Fig. 9.11 shows. This is not a phenomenon that can be predicted by any of the models we have considered so far, and indeed it indicates the presence of acoustic behavior in the airways [163].

The models considered above are all lumped-parameter models described by ordinary differential equations. Such models are appropriate when the flow entering an airway at one end can be considered equal at all times to the flow leaving it at the other end, or when the pressure within an alveolar region of the lung can be considered equal everywhere within that region. These assumptions rely on the transmission of sound across the lung being very rapid compared to the dynamics of the events of interest or, equivalently, the wavelength of sound in air being much larger than the dimensions of

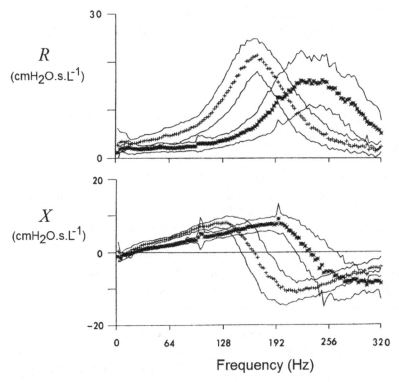

Figure 9.11 High-frequency input impedance in humans when breathing air (+) and a helium-oxygen mixture (×) (reproduced with permission from [178]).

the lung itself. The speed of sound in air is about 300 m.s^{-1}, while the adult human lung has a height around 0.3 m. Consequently, it is not until the frequency reaches well over 100 Hz that the wavelength of sound begins to approach the dimensions of the lung.

The resonances evident between 100 and 300 Hz in Fig. 9.11 reflect the effective quarter-wave resonance of the airway tree. This is the same phenomenon that produces a tone when you blow across the top of an empty soft drink bottle. A resonance is set up in the column of air, in which the air molecules at the bottom are stationary (because they are limited in their movement by the lung tissue, or the bottom of the bottle) while the molecules are the top (the airway opening, or neck of the bottle) are free to vibrate maximally. For a transverse sound wave this represents one quarter of a wavelength. The wavelength of sound at 180 Hz (the resonance for air breathing in Fig. 9.11) is $300/180 = 1.67$ m, yielding a quarter wavelength of 42 cm. This is close to the length one would expect of the combined airway tree, upper airway canal, and any tubing connecting an oscillator to the mouth. The speed of sound in a gas is inversely proportional to the square root of the gas density, so a lighter gas has a higher sound speed. Consequently, the quarter-wave resonance of the airways occurs at a higher frequency with a lighter gas like helium-oxygen compared to air, as seen in Fig. 9.11. This has been modeled by assuming that the airways behave like an acoustic transmission line

[45, 163, 179, 180]. Such models also indicate that the airway wall plays a significant role in the response of the lungs to high-frequency oscillations [166].

Problems

9.1   Suppose that the lungs of two individuals have identical shape, but one lung is half the size of the other in terms of linear dimension. How does the resonant frequency of the smaller lung compare to that of the larger?

9.2   Derive the equation of motion of the regional lung model shown in Fig. 9.5A. Show that its impedance is given by Eq. 9.14. Show that this equation reduces to that of the single-compartment linear model as R_A tends to infinity. Show that as frequency tends to infinity, the resistance of the model in Fig. 9.5A tends to R_{hole}.

9.3   In analogy with Eqs. 9.18 and 9.19, which apply to the viscoelastic model, derive expressions for $R(\omega)$ and $E(\omega)$ for the parallel and series two-compartment models and show what the real and imaginary parts of these expressions tend to as frequency tends to both zero and infinity.

9.4   Extend the six-element model shown in Fig. 9.9 to include a second alveolar compartment in parallel with the distal compartment already in the model. Draw an electrical analog of the model, similar to Fig. 9.9A. Draw a block impedance representation as in Fig. 9.9B.

9.5   In analogy with Eq. 9.27, derive the expression for the transfer impedance of the two-compartment series model (Fig. 7.3).

9.6   Estimate the frequency of the quarter-wave acoustic resonance for air in the lungs of a mouse and an elephant.

10　Constant phase model of impedance

The various lung models we considered in the previous chapter are all composed of collections of discrete elements, each of which is a resistance, an elastance, or a mass. Such models assume that the dissipative, elastic, and inertive properties of the lung are each lumped together in separate physical locations. Accordingly, these models are known as *lumped-parameter models*. Most models of lung mechanics that have appeared in the literature over the past century or so have been of this form. The main reason for the prevalence of lumped-parameter models is that they are described by tractable, if sometimes algebraically tortuous, ordinary differential equations. Also, we tend to be comfortable with the idea of associating individual constitutive properties with distinct components in a model. This tendency to make lumped-parameter models may be cultural; probably many of us can remember learning elementary physics at school with the aid of demonstrations of things like weights suspended on springs. It may also reflect an innate need for the human mind to compartmentalize phenomena in order to make sense of a complex world. In any case, lumped-parameter models lead to a rather artificial view of the way the world actually works, and it is now time to revise this view with respect to the modeling of lung mechanics.

10.1　Genesis of the constant phase model

Our affinity for linear ordinary differential equations instinctively makes us expect to see exponential-type transient responses in nature. Thus, it seemed natural that the first approximation to the volume-time profile of the lung during a relaxed expiration should be a decreasing exponential function of time (Section 7.1). When this was found not to be entirely satisfactory, trying to fit a sum of two exponentials seemed the obvious next step. This led to the spring-and-dashpot Kelvin body (Fig. 7.5) as a mechanical analog of the viscoelastic properties of lung tissue. Spring-and-dashpot models seem to be intuitively reasonable representations of reality, and they can be easily extended simply by adding more Maxwell bodies in parallel, each having its own particular time-constant (Fig. 10.1). Modeling stress relaxation as a sum of exponentials is only useful, however, if the number of exponentials is small, because only then are we dealing with a manageable number of Maxwell bodies. This turns out not to be the case for lung tissue.

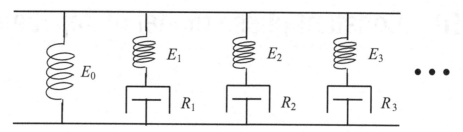

Figure 10.1 Enough Maxwell bodies, each with its own relaxation time-constant (R_i/E_i), arranged in parallel, can mimic any stress adaptation response.

10.1.1   Power-law stress relaxation

If a strip of degassed lung tissue is subjected to a step increase in length (after an appropriate period of pre-conditioning), the resulting sudden increase in tension, T, decreases with a time course that is very accurately described by a temporal *power law* [181]. That is,

$$T(t) = T_0 t^{-k} \tag{10.1}$$

where k and T_0 are constants. Now, in principle it is possible to mimic any conceivable stress relaxation profile with a suitable collection of Maxwell bodies, each having its own time-constant (Fig. 10.1). The problem is, it takes an infinite number of Maxwell bodies to exhibit power-law stress relaxation [96]. This goes against the notion of *parsimony*, one of the universal guiding principles of mathematical modeling, that favors the smallest number of free parameters wherever possible. The experimental appearance of Eq. 10.1 is thus an indication that something may be fundamentally wrong with Fig. 10.1 as an approach to modeling lung tissue mechanics.

Stress relaxation experiments of this kind in isolated whole lungs have found pressure to decrease with the logarithm of time following a sudden increase in volume [182], but this is merely a first-order approximation to the power law because

$$
\begin{aligned}
P(t) &= P_0 t^{-k} \\
&= P_0 (e^{\ln(t)})^{-k} \\
&= P_0 \left[1 - k\ln(t) + \frac{(k\ln(t))^2}{2!} - \cdots \right],
\end{aligned} \tag{10.2}
$$

which is very close to being linear with $\ln(t)$ because k is small (considerably less than 1).

Another important general observation about lung tissue is that the real part of tissue impedance bears an almost fixed relationship to the imaginary part over a wide range of oscillation frequencies. This ratio, termed *hysteresivity* (η), is defined as [183]

$$\eta = \frac{\omega R_t}{E_L} \tag{10.3}$$

and has been measured in numerous species [184, 185] where it typically has a value in the region of 0.1–0.2. In isolated strips of lung tissue, the value of η is somewhat less

[186–188]. η is even fairly well preserved when the lungs are treated with smooth muscle agonists [186]. In other words, it does not seem possible to separately manipulate the resistance and elastance of lung tissue; when one is affected by an intervention, the other invariably also changes. This argues against a lumped-parameter model for lung tissue, and seems to suggest instead that those structures in tissue that store energy elastically are coupled at some fundamental level to those that dissipate it [183].

Power-law stress relaxation (Eq. 10.1) and the frequency independence of η (Eq. 10.3) are actually two facets of the same underlying phenomenon. A power law in pressure describes the response of the lung to a step change in volume. Recalling the linear systems theory outlined in Chapter 8, this means that a power law is also the response to an impulse in flow. Equation 10.1, with T replaced by P, is thus the flow impulse response of the lung. Accordingly, the Fourier transform of Eq. 10.1 is the impedance of the lung (Z), and is found in standard tables to be

$$Z(\omega) = F\{P_0 t^{-k}\}$$

$$= \left(P_0 \sqrt{\frac{2}{\pi}}\right) \Gamma(1-k) \left\{ \frac{\cos\left[(1-k)\frac{\pi}{2}\right] - i\sin\left[(1-k)\frac{\pi}{2}\right]}{\omega^{1-k}} \right\}$$

$$= \frac{G - iH}{\omega^\alpha} \tag{10.4}$$

where Γ is the Gamma function and

$$G = \left(P_0 \sqrt{\frac{2}{\pi}}\right) \Gamma(1-k) \cos\left[(1-k)\frac{\pi}{2}\right], \tag{10.5}$$

$$H = \left(P_0 \sqrt{\frac{2}{\pi}}\right) \Gamma(1-k) \sin\left[(1-k)\frac{\pi}{2}\right], \tag{10.6}$$

$$\alpha = 1 - k, \tag{10.7}$$

and hence

$$\alpha = \frac{2}{\pi} \tan^{-1} \frac{H}{G}. \tag{10.8}$$

Equations 10.5 and 10.6 show where the term *constant phase model* comes from [189]; the ratio of the imaginary to the real part of Z is the tangent of the phase (ϕ) of impedance. That is,

$$\phi = \tan^{-1} \frac{H}{G}$$

$$= (1-k)\frac{\pi}{2}, \tag{10.9}$$

which is independent of frequency and is therefore a constant.

The constant phase model of impedance represents a fundamentally different approach to accounting for tissue mechanics than the model illustrated in Fig. 10.1. To mimic power-law stress relaxation, the model in Fig. 10.1 requires an infinite number of Maxwell bodies [190], so its differential equation of motion contains an infinite number

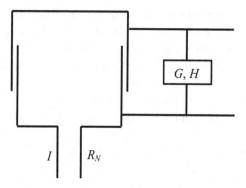

Figure 10.2 The single-compartment lung model with constant phase tissue impedance characterized by the constants G and H. R_N is a Newtonian flow resistance, and I is an inertance.

of free parameters. By contrast, the constant phase model has an impedance (Eq. 10.4) containing only two free parameters, G and H (α is determined by G and H through Eq. 10.8).

10.1.2   Fitting the constant phase model to lung impedance

The viscoelastic model of the lung shown in Fig. 7.5 thus needs to be replaced by a model in which the Kelvin body has been replaced by a construct with an impedance characterized by the parameters G and H (Fig. 10.2). The new model is often referred to in its entirety as the constant phase model, even though the constant phase attribute applies only to the tissue compartment. The resistance leading into this compartment represents a conventional Newtonian resistance (one that dissipates energy in phase with the flow through it) and is denoted R_N. This conduit also has an inertive component I.

The input impedance of the constant phase model of the lung is [157, 189]

$$Z(f) = R_N + i\omega I + \frac{G - iH}{\omega^\alpha}. \tag{10.10}$$

The constant phase model not only describes the impedance of the lung more accurately than the Kelvin body model (Fig. 7.5), it does so with one fewer free parameter. These factors have made Eq. 10.10 the model of choice for describing pulmonary input impedance below about 20 Hz in all species in which it has been examined. Figure 10.3 shows examples from rats and a mouse.

Inspection of Eq. 10.10 shows that the parameters R_N, I, G, and H are all linearly related to Z. Consequently, fitting this model to experimental measurements of Z would be a straightforward matter of applying multiple linear regression were it not for the presence of α in the denominator of the last term. A simple iterative scheme deals with this as follows. Begin by setting $\alpha = 1$. Of course, α is not actually equal to 1, but it is pretty close, so with this initial approximation we can obtain values of R_N, I, G, and H using multiple linear regression. The next step is to calculate a new estimate of α using the estimated values of G and H via Eq. 10.8. Using this new α value in Eq. 10.10, updated estimates of R_N, I, G, and H are obtained, again using multiple linear regression.

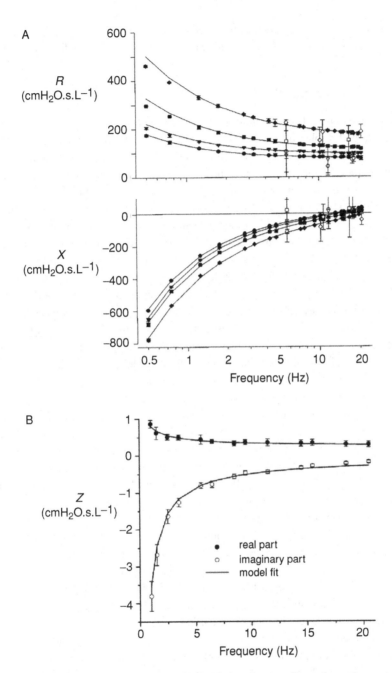

Figure 10.3 (A) Real (top) and imaginary (bottom) parts of impedance from rats under control conditions and at increasing concentrations of intravenous methacholine infusion. The open symbols are data with poor coherence due to interference by the heartbeat. The lines are the fits provided by the constant phase model (reproduced with permission from [191]). (B) Input impedance obtained from a normal, anesthetized, paralyzed and tracheostomized mouse (symbols) together with the fit provided by the constant phase model in Eq. 10.10 (adapted with permission from [170]).

This leads to a further updated estimate of α, and so on. This iterative scheme converges very rapidly (typically to several significant figures within four or five iterations).

10.1.3   Physiological interpretation

Meeting the constant phase model for the first time can be somewhat bewildering. It is natural to be troubled by the question of what G and H represent physically, especially since the Rs and Es of the Maxwell bodies in Fig. 10.1 appeal directly to common intuition. Indeed, one can even imagine constructing a working model based on Fig. 10.1 out of real springs and pistons in cylinders. By contrast, no obvious physical analogs exist for G and H. This relative comfort with spring-and-dashpot models is, however, nothing but an illusion. Lung tissue is a highly complex system of cells, fibers, biomolecules, and fluid from which the bulk elastic and dissipative properties of the tissue somehow arise. A number of attempts have been made to predict tissue mechanics as an emergent phenomenon arising from the ensemble behavior of its components [99, 181, 190, 192–194]. Even so, the precise manner in which this happens remains poorly understood. What is clear, however, is that there are no definable structures within lung tissue that can be identified with springs and dashpots any more than they can with G and H. The spring-and-dashpot and constant phase models are thus on an equal footing in terms of their degrees of empiricism. The constant phase model is, however, vastly superior in terms of parsimony.

R_N in Eq. 10.10 is designated a Newtonian resistance on account of its empirical nature. However, in an animal with an open chest (i.e. when $Z(\omega)$ pertains to the lungs alone), R_N provides an accurate measure of the overall resistance of the airway tree. This has been demonstrated by experiments with the alveolar capsule showing that, on average, the resistance between the trachea and the sub-pleural alveoli is the same as that estimated by R_N [16, 165]. Figure 10.4 shows evidence for this in mice. Correspondingly, I represents the inertance of the gas in the central airways. R_N and I together thus account for the impedance of the airways.

The impedance of the tissues is accounted for collectively by G and H. G is commonly called *tissue damping* and represents the dissipative component of this impedance. G is therefore related to tissue resistance (R_t). Note, however, that G is not exactly the same as R_t because G does not have units of resistance. The same applies to H, which characterizes elastic energy storage within the tissues, but again does not have the same units as elastance. In fact, G and H have the curious feature of having units that are determined by their relative values. Inspection of Eq. 10.10 shows that when Z is expressed, for example, in units of $cmH_2O.s.L^{-1}$, then G and H each have units of $cmH_2O.s^{1-\alpha}.L^{-1}$. These units change with the value of α, which in turn is determined by the ratio H/G (Eq. 10.8).

An elegant way of sidestepping this philosophically troubling issue of variable units is to normalize the tissue impedance term in Eq. 10.10 with respect to a reference frequency ω_0 such that [195]

$$Z(\omega) = R_N + i(\omega)I + \frac{G - iH}{(\omega/\omega_0)^\alpha}. \tag{10.11}$$

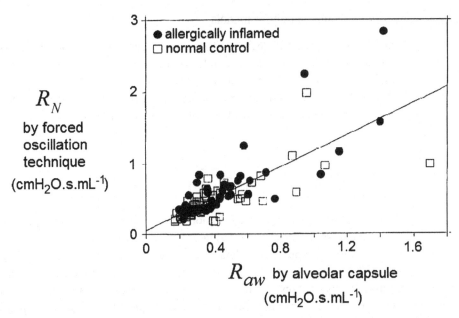

Figure 10.4 Estimates of R_{aw} made directly by alveolar capsule versus R_N obtained by fitting Eq. 10.10 to impedance measurements in mice (adapted with permission from [16]).

By choosing $\omega_0 = 1$, the numerical values of G and H remain unchanged, yet they now have the same units as Z itself. In particular, G has units of resistance and can be labeled as such. On the other hand, H also has units of impedance, which might seem somewhat unsatisfactory given its obvious functional relationship to elastance. Nevertheless, this is still better than having variable units. Alternatively, even without the above mathematical sleight of hand involving ω_0, one can still equate H to the value of conventional elastance obtained at an oscillation frequency of $\omega = 1$. At this frequency, the imaginary parts of the tissue impedances of the constant phase model (Eq. 10.10) and single compartment model (Eq. 9.7) become equal to each other.

10.2   Heterogeneity and the constant phase model

The constant phase model assumes that the lung is ventilated homogeneously, so that the pressures in all alveoli at any given time are identical. While this seems to be a good approximation in a normal lung over the frequency range of spontaneous breathing, it clearly breaks down as pathology sets in or when the lungs constrict in response to a bronchial agonist. Indeed, having to deal with heterogeneities provided the prime motivation for progressing beyond two-compartment models to the general linear model covered in Chapter 8. How, then, can the constant phase model, being based on the assumption of homogeneous ventilation, deal with ventilation heterogeneity throughout the lung?

10.2.1   Distributed constant phase models

One obvious approach to dealing with regional variations in lung mechanics is to model
them explicitly by connecting multiple constant phase compartments together. This is
an obvious extension of the philosophy that led from the single-compartment model in
Chapter 3 to the two-compartment models considered in Chapter 7. N constant phase
models connected in parallel lead to the impedance expression

$$Z(\omega) = \left[\sum_{k=1}^{N} \frac{1}{R_{N_k} + i\omega I_k + \dfrac{G_k - iH_k}{\omega^{\alpha_k}}} \right]^{-1}.$$
(10.12)

As N increases, however, this expression rapidly leads to an unmanageable number of
free parameters. This makes the fitting of Eq. 10.12 to data untenable.

An alternative approach is to imagine the lung to be composed of a very large number
of parallel constant phase compartments, but have the parameters of the compartments
conform to distributions that themselves are characterized by a small number of parameters. Additional constraints help to further simplify the situation. In particular, the
experimentally observed tendency for η to remain constant in the face of changing
experimental conditions suggests that it may be reasonable to assume η is the same for
all compartments. If one further assumes that regional heterogeneity of ventilation is
due solely to variations in local tissue properties then the values of R_N and I are also the
same for all compartments. This leaves H as the only quantity that varies in the model,
so the problem is reduced to one of choosing a probability distribution $\xi(H)$ from which
H is to be drawn [195]. Equation 10.12 then reduces to

$$Z(\omega) = \left[\int_0^\infty \frac{\xi(H)}{R_N + i\omega I + \dfrac{(\eta - i)H}{\omega^\alpha}} d\xi \right]^{-1}$$
(10.13)

where the area under $H(\xi)$ is unity. If, for example, $H(\xi)$ is a truncated hyperbolic distribution having lower and upper extremes $H_{\min}$ and $H_{\max}$, respectively, then
Eq. 10.13 becomes [195]

$$Z(\omega) = \left[\int_{H_{\min}}^{H_{\max}} \frac{\dfrac{1}{\ln\left(\dfrac{H_{\max}}{H_{\min}}\right)H}}{R_N + i\omega I + \dfrac{(\eta - i)H}{\omega^\alpha}} d\xi \right]^{-1}$$

$$= \frac{\ln\left(\dfrac{H_{\max}}{H_{\min}}\right)(R + i\omega I)}{\ln\left(\dfrac{H_{\max}}{H_{\min}}\right) + \ln\left(\dfrac{(\eta - i)H_{\min} + \omega^\alpha(R + i\omega I)}{(\eta - i)H_{\max} + \omega^\alpha(R + i\omega I)}\right)}.$$
(10.14)

This model has been found to give an improved fit to measurements of $Z(\omega)$ obtained in mice following treatment with elastase, which produces some of the pathological features of emphysema [195]. A similar model in which R_N, instead of H, varies across the various compartments has been shown to give improved fits to impedance data from mice during methacholine-induced bronchoconstriction [196]. The distributed model has also been applied to lung impedance during acute lung injury [106, 197].

These distributed-parameter versions of the constant phase model, however, encounter inevitable problems with model uniqueness. It is not possible to determine from measurements of input impedance alone whether the lung is better described by a distribution of different values for R or for E. Furthermore, the lungs may become compartmentalized in a serial fashion as well as in parallel, particularly during severe narrowing of distal airways, which causes imposed oscillations in flow to be shunted to the compliant central airways. Consequently, it is not always easy to judge if the R or H distributions recovered using distributed constant phase models are really reflective of the inner workings of the lung, or simply empirical functions that merely permit a good fit to the data.

10.2.2   Heterogeneity and hysteresivity

An alternative approach to acknowledging the presence of heterogeneity with the constant phase model arises from the experimental observation that η always seems to increase during bronchoconstriction [54, 198, 199]. Alveolar capsule experiments have also shown that heterogeneity increases as bronchoconstriction increases in severity [74, 133, 134]. This has led to the notion that η always increases as the lung becomes mechanically heterogeneous, and therefore can be taken as an indicator of heterogeneity. Support for this notion has been provided by studies using anatomically based forward models of the lung in which Eq. 10.10 has been fit to impedances simulated under conditions of known heterogeneity [46–48]. These studies have shown that η increases progressively with the degree of heterogeneity present in the forward model.

We thus have a somewhat curious situation; the presence of heterogeneity is being inferred from the way that parameter values change in a model which itself is specifically incapable of representing heterogeneity. This is nevertheless a convenient state of affairs because it means we can monitor heterogeneity without having to deal with more complicated models, such as that expressed in Eq. 10.13, and their attendant issues of uniqueness. On the other hand, we have to be careful not to be mislead. Using η as an index of heterogeneity is only useful if η and heterogeneity always (or almost always) increase together. As it turns out, this seems to be the case, although for reasons that have more to do with the physics of the airways rather than mathematical imperative, as the following analysis shows [200].

An analytical evaluation of the effects of heterogeneity on η with regard to the constant phase model is problematic because impedance data simulated over a range of frequencies using a multi-compartment constant phase model cannot be fit exactly by Eq. 10.10. However, η can also be defined as the ratio of lung resistance (E_L) to tissue elastance (R_t) at a single frequency (Fig. 10.3). Indeed, this was the original definition of η [183]. As we saw in Chapter 9, formulae for the effective values of E_L and R_t at a

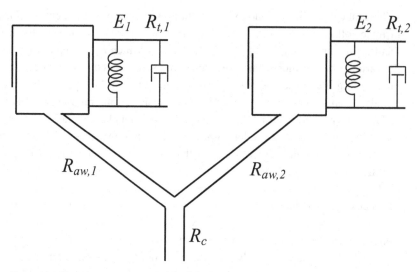

Figure 10.5 Two-compartment model of the lung. Each compartment is represented by a pair of telescoping cylinders linked together by a spring (elastance E_j) and dashpot (resistance $R_{t,j}$), and served by an airway (resistance $R_{aw,j}$). The two compartmental airways communicate with the environment via a common airway (resistance R_c).

particular frequency can be calculated for multi-compartment models of the lung so long as the models are linear. When such a model is heterogeneous (i.e. its compartments have different time-constants) then E_L and R_t depend on the resistances and elastances of the individual components in the model. This is illustrated by Eqs. 9.18 and 9.19.

Consider the two-compartment model shown in Fig. 10.5. The tissue properties of each compartment are characterized by a single elastance E_j and resistance $R_{t,j}$, where $j = 1$ or 2. The compartments each have an airway with resistance $R_{aw,j}$ and communicate with the environment through a common central airway with resistance R_c. The impedance of this model is

$$Z(\omega) = R_c + \frac{\left(R_1 - \dfrac{i E_1}{\omega}\right)\left(R_2 - \dfrac{i E_2}{\omega}\right)}{R_1 + R_2 - \dfrac{i (E_1 + E_2)}{\omega}} \qquad (10.15)$$

where $R_{aw,j} + R_{t,j}$ have been replaced by R_j. The resistance, $R(\omega)$, of the model is the real part of Eq. 10.15, namely

$$R(\omega) = R_c + \frac{\omega^2 R_1 R_2(R_1 + R_2) + R_1 E_2^2 + R_2 E_1^2}{\omega^2 (R_1 + R_2)^2 + (E_1 + E_2)^2}. \qquad (10.16)$$

When the time-constants of the two compartments are equal (i.e. $R_1/E_1 = R_2/E_2$), the model behaves like a single homogeneous compartment and $R(\omega)$ collapses to

$$R_0 = R_c + \frac{R_1 R_2}{R_1 + R_2}. \qquad (10.17)$$

The increase in the resistance of the model due to the presence of heterogeneities is thus

$$\Delta R = R(\omega) - R_0$$

$$= \left[R_c + \frac{\omega^2 R_1 R_2 (R_1 + R_2) + R_1 E_2^2 + R_2 E_1^2}{\omega^2 (R_1 + R_2)^2 + (E_1 + E_2)^2} \right] - \left[R_c + \frac{R_1 R_2}{R_1 + R_2} \right]$$

$$= \frac{(R_1 E_2 - R_2 E_1)^2}{[\omega^2 (R_1 + R_2)^2 + (E_1 + E_2)^2](R_1 + R_2)}. \qquad (10.18)$$

Now, it is known that the overall tissue resistance of the lung (R_t) decreases monotonically with frequency, eventually becoming negligible compared to the resistance of the airway tree (R_{aw}) [135, 157]. Thus, in the limit of large ω we obtain an expression for R_{aw} as

$$R_{aw} = R_c + \frac{R_{aw,1} R_{aw,2}}{R_{aw,1} + R_{aw,2}}. \qquad (10.19)$$

Furthermore, when the lung is homogeneous, ΔR is zero, in which case

$$R_{\text{hom}} = R_{aw} + R_t \qquad (10.20)$$

where

$$R_t = \frac{R_{t,1} R_{t,2}}{R_{t,1} + R_{t,2}}. \qquad (10.21)$$

In the heterogeneous lung, therefore, R is given by the sum of Eqs. 10.18, 10.19, and 10.21, so the apparent resistance of the tissues is

$$R_t(\omega) = R_t + \Delta R$$

$$= \frac{R_{t,1} R_{t,2}}{R_{t,1} + R_{t,2}} + \frac{(R_1 E_2 - R_2 E_1)^2}{[\omega^2 (R_1 + R_2)^2 + (E_1 + E_2)^2](R_1 + R_2)}. \qquad (10.22)$$

The elastance, $E(\omega)$, of the model is the real part of Eq. 10.15 multiplied by $-\omega$. That is,

$$E(\omega) = \frac{E_1 E_2 (E_1 + E_2) + \omega^2 \left(E_1 R_2^2 + E_2 R_1^2 \right)}{\omega^2 (R_1 + R_2)^2 + (E_1 + E_2)^2}. \qquad (10.23)$$

Like $R(\omega)$, $E(\omega)$ has a minimum (E_0) in the homogeneous case when the time-constants of the two compartments are equal. That is,

$$E_0 = \frac{E_1 E_2}{E_1 + E_2}. \qquad (10.24)$$

The increase in $E(\omega)$ due to the presence of heterogeneities is thus

$$\Delta E = E - E_0$$

$$= \frac{\omega^2 (R_1 E_2 - R_2 E_1)^2}{(E_1 + E_2) \left[\omega^2 (R_1 + R_2)^2 + (E_1 + E_2)^2 \right]}. \qquad (10.25)$$

In other words,

$$E(\omega) = E_0 + \frac{\omega^2 (R_1 E_2 - R_2 E_1)^2}{(E_1 + E_2) \left[\omega^2 (R_1 + R_2)^2 + (E_1 + E_2)^2 \right]}. \qquad (10.26)$$

Using Eqs. 10.22 and 10.26 we obtain

$$\eta = \omega \frac{R_t(\omega)}{E(\omega)}$$

$$= \frac{\omega R_t + \dfrac{\omega (R_1 E_2 - R_2 E_1)^2}{(R_1 + R_2)[\omega^2 (R_1 + R_2)^2 + (E_1 + E_2)^2]}}{E_0 + \dfrac{\omega^2 (R_1 E_2 - R_2 E_1)^2}{(E_1 + E_2)[\omega^2 (R_1 + R_2)^2 + (E_1 + E_2)^2]}}. \tag{10.27}$$

As can be seen immediately, when the lung is homogeneous (i.e. the time-constants of the two compartments are equal so that $R_1 E_2 = R_2 E_1$), this expression for η reduces to simply $\omega R_t / E_0$. The question of whether heterogeneity increases η thus comes down to establishing the condition under which the inequality

$$\frac{\omega R_t + \dfrac{\omega (R_1 E_2 - R_2 E_1)^2}{(R_1 + R_2)[\omega^2 (R_1 + R_2)^2 + (E_1 + E_2)^2]}}{E_0 + \dfrac{\omega^2 (R_1 E_2 - R_2 E_1)^2}{(E_1 + E_2)[\omega^2 (R_1 + R_2)^2 + (E_1 + E_2)^2]}} > \frac{\omega R_t}{E_0} \tag{10.28}$$

is satisfied. After some algebra, this inequality reduces to

$$\frac{E_0}{R_t} > \frac{\omega^2 (R_1 + R_2)}{E_1 + E_2}. \tag{10.29}$$

Now, because everything in the above inequality has fixed value except ω, this inequality will cease to hold when ω becomes large enough, regardless of the values of the other parameters. This means that the inequality is not true in general, so an increase in heterogeneity does *not* automatically mean that η will also increase. However, ω is close to unity over the frequency range of normal breathing. Furthermore, airway and tissue resistances in a normal lung are numerically about an order of magnitude smaller than tissue elastance [15]. Therefore, when oscillation frequency is not too high, and the lung is normal or mildly heterogeneous, one expects the above inequality to be satisfied. This may explain why an increase in η with the onset of heterogeneity has been so widely reported.

On the other hand, as airway constriction increases in severity, one would think it should be possible for the R_i to achieve arbitrarily high values, eventually violating the above inequality. However, a decreasing value of η in the presence of severe bronchoconstriction does not seem to be observed, which suggests that airways do not become arbitrarily narrow, even though some airways have been shown to close completely. An interpretation of this apparent conundrum is that when airways narrow to a certain point they make a rapid transition to complete closure, perhaps due to the sudden formation of liquid bridges across the lumen [47, 91, 170].

The above reasoning represents an interesting and somewhat unusual way to use an inverse model to link structure to function. The diagnostic capability of the constant phase model can be extended through the use of *a priori* information about the behavior of its parameters, specifically the ratio G/H, without actually extending the model itself. Of course, one has to be careful. There are situations where an increase in η can be produced by phenomena other than heterogeneity. It can be shown, for example, that recruitment and derecruitment of lung units can cause η to either increase or decrease, depending on the details of the situation [200]. Nevertheless, when the onset of heterogeneities is expected, an increase in η can serve as a useful indicator for their presence.

10.3   The fractional calculus

In Chapter 8 we saw how the impedance of a compartment model of the lung can be determined by taking the Fourier transform of its differential equation of motion. We now consider how to do this for the constant phase model. First, however, we have to figure out the differential equation that describes the constant phase model. One possibility is suggested by Fig. 10.1. That is, determine the first-order equations that relate the pressures across each Maxwell body to the displacements of their springs and dashpots, and then combine these equations into a single master equation containing only the total pressure and flow, as was done for the two-compartment models in Chapter 7. There is a problem here, however. As we discussed above, it takes an infinite number of Maxwell bodies to behave like the constant phase model, so the resulting differential equation will be of infinite order. The simplicity of the constant phase model in the frequency domain thus does not seem to be matched with a corresponding elegance by ordinary differential equations in the time domain.

An alternative approach to the problem of finding a differential equation for the constant phase model is suggested by the fact that differentiation in the time domain is equivalent to multiplication by $i\omega$ in the frequency domain. Indeed, a time-domain derivative of arbitrary order is represented in the frequency domain simply by raising $i\omega$ to the appropriate integer power (Section 9.1). When examined in this regard, however, the constant phase model of tissue impedance takes on a curious form. We begin by expressing the model in terms of amplitude and phase thus:

$$Z_t(\omega) = \frac{G - iH}{\omega^\alpha}$$
$$= Ae^{-i\phi} \tag{10.30}$$

where

$$\phi = \tan^{-1}\frac{H}{G} \tag{10.31}$$

and

$$A = \sqrt{\frac{G^2 + H^2}{\omega^{2\alpha}}}. \tag{10.32}$$

The amplitude and phase expression for $Z_t(\omega)$ corresponds to its representation as a vector in the complex plane (the *Argand diagram*). The length (A) of this vector depends on frequency but its phase (ϕ) does not, as illustrated in Fig. 10.6.

The axes of the Argand diagram can be labeled with integer powers of i. That is, the positive real axis corresponds to i raised to the power 0, the positive imaginary axis i

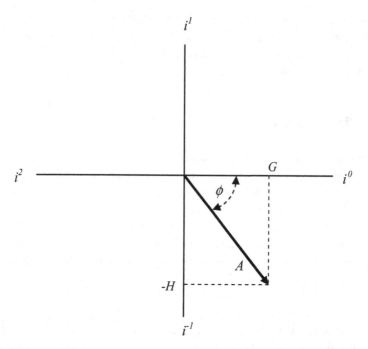

Figure 10.6 Argand diagram representation of Z_t as a vector having frequency-dependent length A and constant phase ϕ.

raised to the power 1, and so on. By analogy, the orientation of the vector $\underline{A}$ at fixed angle ϕ to the horizontal axis can be represented as i raised to a power equal to ϕ as a fraction of $\pi/2$ radians. From Eqs. 10.8 and 10.31, this fraction can be seen to equal the constant α of the constant phase model.

The logical conclusion of all this is that because ordinary differential equations contain factors composed of i raised to integer powers, what we must be dealing with here is a derivative that is neither of order 0 nor order -1, but something in between. Indeed, we can re-write Eq. 10.30 as

$$Z_t(\omega) = \frac{E}{(i\omega)^\alpha},\tag{10.33}$$

which translates back into the time domain to the equation of motion

$$P(t) = E\frac{d^{-\alpha}\dot{V}(t)}{dt^{-\alpha}}.\tag{10.34}$$

Thus, rather than having an ordinary differential equation of infinite degree to describe the constant phase model, we instead have an equation containing a single *fractional derivative* [190]. This is parsimony indeed!

Despite the mathematical elegance of the fractional calculus in this situation, however, we are still left with having to try to interpret its consequences in physical terms. Recall from Chapter 8 that differentiation of order 0 is equivalent to convolution with a delta function. The first-order derivative thus contains only local information about the original function. Conversely, differentiation of order −1 (i.e. integration) is equivalent to convolution with a step function, which has an infinite memory for past values of the original function. Differentiation of order −α, where $0 > -\alpha > -1$, is equivalent to convolution with a hyperbolic function of frequency. That is,

$$\frac{d^{-\alpha} \dot{V}(t)}{dt^{-\alpha}} = \frac{1}{\Gamma(\alpha)} \int_{0}^{t} \dot{V}(u) \frac{1}{(t-u)^{1-\alpha}} \, du. \tag{10.35}$$

Thus, as the order of differentiation progresses from 0 to −α, the differentiated function gains a progressively stronger memory for past values of $\dot{V}(t)$.

The fractional calculus has been applied to numerous physical systems that appear to exhibit scale-free dynamics [201–203]. The constant phase nature of lung tissue is scale-free because it reflects mechanical properties that have no preferred time scale. In other words, there is no equivalent of a resonant frequency. In fact, this property seems to characterize biological soft tissues in general, and might be viewed as advantageous given that many tissues in the body are intimately juxtaposed and so have to synchronize their movements. With no tissue having a characteristic frequency, there are no problems of mechanical impedance-matching that arise between different tissues.

10.4   Applications of the constant phase model

The great utility of the constant phase model lies in the fact that, with only four free parameters, it can accurately partition the impedance of the normal lung into a component due to the airways (characterized by R_N and I) and a component due to the tissues (characterized by G and H). This represents a significant advance over the single-compartment model (Eq. 3.5), the resistance of which contains a substantial, but usually unknown, contribution from tissue resistance. Interpreting changes in the parameters of the constant phase model can thus provide physiological insight into how a particular intervention affects the lung. This is most readily achieved with respect to R_N and I, which reflect airway caliber. Bronchoconstriction, for example, causes the airways to narrow, which increases R_N and I (Eq. 9.10) accordingly.

Interpreting changes in G and H is not quite as straightforward. There are several different possible ways to explain changes in G and H, one of which is that the intrinsic rheological properties of the tissues have changed. Alternatively, if a fraction of the lung becomes closed off from the airway opening, as a result of either airway closure or alveolar collapse, then both G and H will increase by the same fraction. Pure derecruitment thus preserves η. On the other hand, as we have seen (Section 10.2.2), if the lung becomes heterogeneous then G will very likely increase proportionately more than H.

All of these various facets of parameter behavior are evident in the data shown in Fig. 10.7, which shows time profiles of R_N, G, and H during the five minutes following exposure of mice to an aerosol of the bronchial agonist methacholine. (I makes a negligible contribution to impedance in mice below 20 Hz because the masses of gas involved are so small.) The response begins with sharply defined peaks in R_N and G that

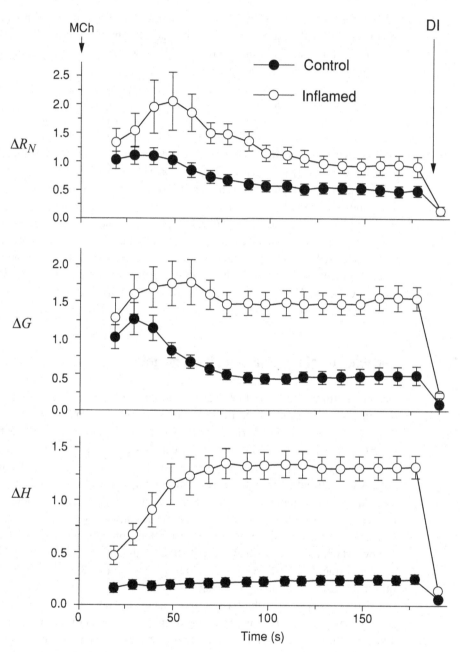

Figure 10.7 The parameters of respiratory input impedance (R_N, G, and H) expressed as fractional changes above baseline (ΔR_N, ΔG, and ΔH, respectively) in control and allergically inflamed mice following a 40-second challenge with an aerosol of methacholine. MCh indicates time of completion of delivery of methacholine. DI indicates time of delivery of two deep lung inflations to 25 cmH$_2$O (reproduced with permission from [170]).

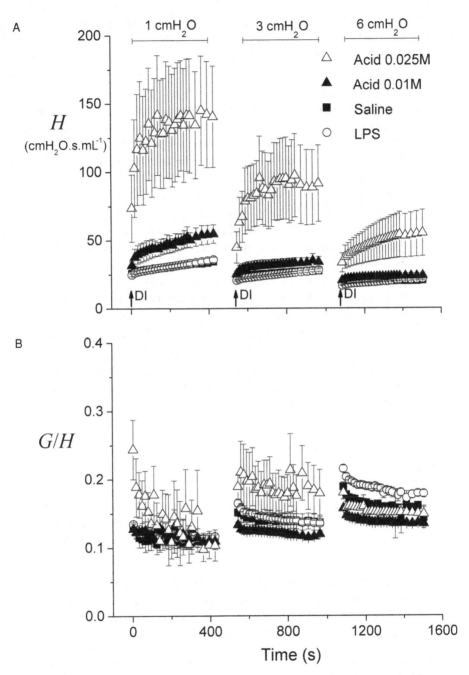

Figure 10.8 (A) H versus time at PEEP levels of 1, 3, and 6 cmH$_2$O for mice treated with lipopolysaccharide (LPS), two different concentrations of hydrochloric acid, and saline as a control. A deep inflation (DI) of the lungs was administered at the beginning of each run (indicated by vertical arrows). The data shown are mean values ± one standard error, illustrating the wide variability in the degrees of lung injury manifest in different animals. (B) Corresponding plots of hysteresivity (G/H) (reproduced with permission from [92]).

resolve to elevated plateaus. H, by contrast, exhibits only the plateau phase. In allergically inflamed mice these responses are quantitatively very different (open circles in Fig. 10.8), but qualitatively the broad features are the same. Simulations with an anatomically based computational model of the mouse lung indicate the following interpretation of these data [170]. The initial peaks in R and G reflect active contraction followed by relaxation of the airway smooth muscle. The plateaus represent an incomplete return to baseline, and in the case of H quantify the degree of airway closure caused by the bronchoconstriction. This closure is not reversed until the lungs are given a deep breath to reopen the airways. The fact that the closure was much more pronounced in the inflamed mice leads to the conclusion that the allergic mice used in this study were hyperresponsive entirely because they had a physically thickened and more mucus-laden epithelium. This interpretation has been supported by micro-computed tomography imaging of mouse lungs showing areas of complete collapse behind closed airways [26].

In contrast to bronchoconstriction, acute lung injury is a pathology that affects the lung tissues more than the airways. One would therefore expect to see a larger effect on G and H than on R_N. This indeed turns out to be the case, as Fig. 10.8 illustrates. Here are shown plots of H versus time when the lungs were given a deep inflation to recruit as much of the lung as possible. Then mechanical ventilation with a normal tidal volume was delivered for the next five minutes [70]. H increased slightly over this period in normal animals, but in animals with severe acute lung injury the increase was rapid and dramatic (Fig. 10.8, top panel). Corresponding plots of η (i.e. G/H), however, show relatively little change (Fig. 10.8, bottom panel). This suggests that the increases in H occurring over time were due purely to progressive derecruitment of lung units.

Problems

10.1   Equation 10.1 gives the force response of a strip of lung tissue to a step in length. What is the expression for the impulse response? Why is it problematic to express a general force profile for the tissue by convolving a length profile with this impulse response?

10.2   The ratio of G to H (the two tissue parameters in the constant phase model) does not change with frequency. What is the mathematical expression for this ratio, termed hysteresivity, for the viscoelastic model shown in Fig. 7.5? Sketch a plot of hysteresivity for the viscoelastic model as a function of frequency.

10.3   In analogy with Eqs. 10.12 to 10.14, derive an expression for the impedance of a model consisting of a continuous distribution of parallel constant phase models in which the value of R_N varies while H remains the same for all compartments.

10.4   Prove that E_0 given by Eq. 10.24 is the minimum value of $E(\omega)$ given by Eq. 10.23 when E_1 and E_2 are fixed and R_1 and R_2 are allowed to vary.

10.5   Show that the inequalities in Eqs. 10.28 and 10.29 are equivalent.

10.6   It sometimes happens that fitting the constant phase model to $Z(\omega)$ data obtained from mechanically heterogeneous lungs produces negative values for R_N. How might this happen? Illustrate your answer with a sketch. (Hint: R_N and the other parameters of the constant phase model are obtained by fitting the model equation to impedance obtained over a finite range of frequency, yet R_N is the value of the real part of impedance at infinite frequency.)

10.7   How would you expect the parameters of the constant phase model (R_N, I, G, and H) to vary with lung volume?

10.8   Explain how Eq. 10.35 becomes equivalent to integration as α tends to 1.

11   Nonlinear dynamic models

We are now ready to tackle the fourth and final level of complexity in inverse models of the lung, that of the nonlinear dynamic model. Here, the sky is the limit in terms of generality, but unfortunately complexity increases accordingly. We will begin by examining a general theory of dynamic nonlinear systems and then investigate how special cases of this general theory have been applied to the description of lung mechanics.

11.1   Theory of nonlinear systems

11.1.1   The Volterra series

For causal systems that have asymptotically finite memory (which does not include systems with static hysteresis that remember past histories forever), the general nonlinear dynamic model is described by the *Volterra series* [204]. This links an output $y(t)$ to an input $x(t)$ through an infinite sequence of terms. To give an idea of just how bad things can get, here are the first four terms of the series:

$$
y(t) = h_0 + \int_0^\infty h_1(u_1)x(t - u_1)\,du_1
$$

$$
+ \int_0^\infty \int_0^\infty h_2(u_1, u_2)x(t - u_1)x(t - u_2)\,du_1du_2
$$

$$
+ \int_0^\infty \int_0^\infty \int_0^\infty h_3(u_1, u_2, u_3)x(t - u_1)x(t - u_2)x(t - u_3)\,du_1du_2du_3 + \cdots \quad (11.1)
$$

where the u_i are dummy variables of integration, each representing time, and the functions h_1, h_2, h_3, etc. are called *kernels*.

The first-order term in Eq. 11.1 is the convolution integral we met in Chapter 8 (Eq. 8.14), so the Volterra series to first order is simply the linear dynamic system that describes the present output as a weighted sum of past inputs. The first-order kernel $h_1(t)$ is the impulse response, and the identification of a linear system boils down to determining only this first-order kernel because all higher-order kernels are zero. The second-order Volterra term contains a two-dimensional kernel $h_2(t_1, t_2)$ that accounts

Wiener model

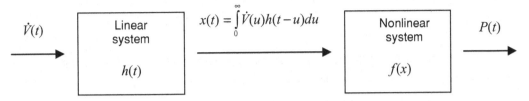

Hammerstein model

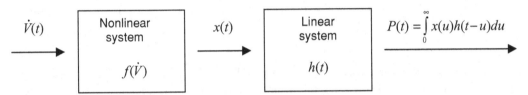

Figure 11.1 Wiener and Hammerstein models of lung mechanics showing how an input $\dot{V}$ signal is converted into an output P signal. h is the impulse response function of a linear system. f is a static nonlinear function.

for the way in which inputs at two past times $t - u_1$ and $t - u_2$ interact to influence the current output. The third-order kernel is a three-dimensional function, and so on. Identification of the general nonlinear dynamic system thus involves determining an infinite series of terms of ever-increasing dimensionality. This is a problem of incomprehensible magnitude, particularly when it comes to making physical sense of all the kernels. Consequently, highly reduced versions of the Volterra series are employed for the description of real systems, such as that used to model dynamic pressure-volume loops from dog lungs [205].

11.1.2   Block-structured nonlinear models

A useful class of nonlinear model consists of the *block-structured* constructs that include the *Wiener* and *Hammerstein* models, which have been widely used to model numerous systems including the lung [206–209]. These models have the particular attribute that the static and dynamic aspects of their behavior are separable. This means that these static and dynamic attributes can be modeled as occurring within distinct blocks linked together in series. Furthermore, the dynamic block is modeled as a linear system while all nonlinear behavior is confined to the static block. In the Wiener model, the input feeds into the linear dynamic system, the output of which is then acted upon by a static nonlinearity (Fig. 11.1A). In the Hammerstein model, the order of the blocks is reversed.

The Wiener and Hammerstein models are special cases of the general Volterra model, as the following analysis shows. In the case of the Wiener model, the input $\dot{V}$ is first converted to an intermediate variable x via the convolution integral

$$x(t) = \int_0^\infty h(u)\dot{V}(t-u)\,du. \tag{11.2}$$

This is then subjected to a nonlinear transformation. Such a transformation can be expressed as the polynomial series

$$P(\dot{V}) = a_0 + \sum_{i=1}^\infty a_i \left[\int_0^\infty h(u)\dot{V}(t-u)\,du \right]^i$$

$$= a_0 + a_1 \int_0^\infty h(u)\dot{V}(t-u)\,du + a_2 \int_0^\infty h(u)\dot{V}(t-u)\,du \int_0^\infty h(u)\dot{V}(t-u)\,du$$

$$+ \cdots \tag{11.3}$$

where the a_i are constants. Comparison of Eq. 11.1 and Eq. 11.3 shows that the higher-order Volterra kernels for the Wiener model are simply products of the first-order kernel.

The relationship of the Hammerstein model to the Volterra series can be derived in a similar fashion. In the Hammerstein model, the nonlinear function f operates on the input $\dot{V}$, which can be expressed as the polynomial series

$$f(\dot{V}) = a_0 + \sum_{i=1}^\infty a_i V^i. \tag{11.4}$$

Substitution of Eq. 11.4 into the convolution integral describing the subsequent linear system gives

$$P(t) = a_0 + \sum_{i=1}^\infty a_i \int_0^\infty h(u)V^i(t-u)\,du$$

$$= a_0 + a_1 \int_0^\infty h(u)\dot{V}(t-u)\,du + a_2 \int_0^\infty h(u)\dot{V}(t-u)\dot{V}(t-u)\,du + \cdots. \tag{11.5}$$

Comparison of Eqs. 11.1 and 11.5 reveals that the higher-order kernels of the Hammerstein model are simply scaled versions of the first-order kernel and therefore have non-zero values only on the leading diagonal.

11.2   Nonlinear system identification

Nonlinear system identification is the process of fitting nonlinear dynamic models, such as those just described, to experimental data. Nonlinear system identification can be usefully applied to the study of lung mechanics, as we shall see shortly. First, however, we will examine the consequences of performing linear system identification on a nonlinear system because this characterizes any application of the forced oscillation technique described in Chapter 8. Like any real-world system, the lung never behaves in a perfectly linear fashion. Indeed, sometimes the linearity assumption can be badly violated, particularly when oscillation amplitudes are large. It is therefore important to understand how nonlinear mechanical behavior affects the determination of lung impedance.

11.2.1 Harmonic distortion

The Volterra series can be taken into the frequency domain via an extension of the process outlined in Section 8.2.3. This gives

$$Y(\omega) = H_1(\omega)X(\omega) + \int_{-\infty}^{\infty} H_2(\upsilon, \omega - \upsilon)X(\upsilon)X(\omega - \upsilon)\,d\upsilon$$

$$+ \int_{-\infty}^{\infty}\int_{-\infty}^{\infty} H_3(\psi, \upsilon, \omega - \upsilon - \upsilon)X(\psi)X(\upsilon)X(\omega - \psi - \upsilon)\,d\psi\,d\upsilon + \cdots \quad (11.6)$$

where Fourier transforms of time-domain functions are denoted by upper-case letters. The first term of Eq. 11.6 is the familiar product corresponding to a linear system. The second term contains a symmetric two-dimensional kernel H_2 that describes how pairs of different frequency components in the input interact to produce a contribution to the output at the frequency in question. The third term accounts for interactions between triplets of frequencies in the input, and so on.

Equation 11.6 clearly shows how the principle of superposition fails for nonlinear systems. In contrast to the situation for linear systems, in a general nonlinear system each output component can potentially receive contributions from all frequency components in the input. This means that when a nonlinear system is perturbed with a broad-band input, one does not know how much of the output power at any particular frequency came from the input at that frequency and how much came from input at other frequencies.

When the input to a nonlinear system consists of only a single frequency, the output will in general consist of a component at that frequency plus additional components at the harmonics of the input. This can be seen from Eq. 11.6. In the second term of this equation, for example, $X(\upsilon)$ is nonzero only when υ equals the input frequency, so $X(\omega - \upsilon)$ is zero only when ω equals 2υ. This term thus produces a contribution to the output at the first harmonic of the input. By the same reasoning, the third term in Eq. 11.6 produces a contribution to the output at the second harmonic of the input. Accordingly, when the input is a pure sine wave at a given frequency, you know for sure that the output at the same frequency derives only from the input. The remaining higher-frequency components in the output represent what is known as *harmonic distortion*. A useful index of the degree of nonlinearity in a system is the *harmonic distortion index*, defined as the ratio of the output power at frequencies above the fundamental to the total output power [210, 211].

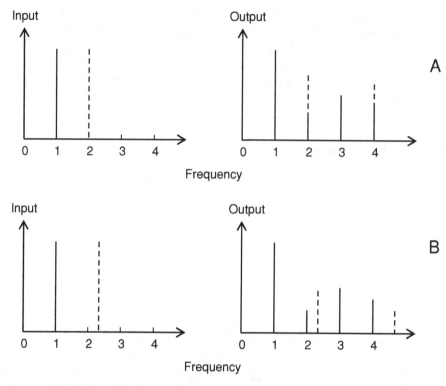

Figure 11.2 (A) When a fundamental frequency (solid line) and its first harmonic (dashed line) are input to a nonlinear dynamic system, the resulting output contains components at those two frequencies together with each of their higher harmonics. Some of the output components may thus be composed of contributions from both input components. (B) When the higher frequency in the input is not an integer multiple of the lower frequency, the harmonics do not overlap. This means that the components in the output at the two input frequencies are derived only from the input components at those respective frequencies.

The above discussion suggests a way to reduce the effects of harmonic distortion on estimates of impedance obtained using the forced oscillation technique. The basic problem is knowing how much of the power at any given frequency in the output comes from the input at that frequency, and how much is harmonic contributions from lower-frequency components in the input. A given frequency in the output will be corrupted by the higher harmonics of lower-frequency input components only if the output frequency is an integer multiple of one or more of the lower input frequencies. This situation can therefore be avoided if the frequencies in the input are *mutually prime*, so that none is an integer multiple of any other [155]. Figure 11.2 illustrates how this works.

Unfortunately, however, this is not the end of the problem. A higher level of harmonic distortion can occur when any frequency in the input is either the sum or difference of any other pair of input frequencies. In this case, a *no-sum-no-difference* input waveform can be used to further reduce harmonic distortion in the output [212].

11.2.2   Identifying Wiener and Hammerstein models

Fitting either the Wiener or Hammerstein model to experimental measurements of $\dot{V}$ and P means determining the impulse response function $h(t)$ and the nonlinear function f (Fig. 11.1) that give the best prediction of P from $\dot{V}$ in the least squares sense. This is achieved iteratively as follows [213]. In the case of the Wiener model, an initial estimate of $h(t)$ is obtained by treating P and $\dot{V}$ as if they are related via a linear system. In other words, $P(t)$ is deconvolved with $\dot{V}(t)$. This can be done using the Fourier techniques outlined in Chapter 8 to estimate an impedance, which is then submitted to the inverse FFT. Next, $\dot{V}$ is reconvolved with the identified $h(t)$ to obtain an estimate of $x(t)$. $P(t)$ is then fit to a polynomial function of the estimated $x(t)$. This polynomial constitutes an initial estimate of f. The algorithm then proceeds to iteratively update these estimates of $h(t)$ and f, in turn, using successive re-estimations of the intermediate function $x(t)$. First, $x(t)$ is re-estimated by submitting P to the inverse of f, and then the result is deconvolved with $\dot{V}$ to provide an update of $h(t)$. Next, the new $h(t)$ is convolved with $\dot{V}$ to further re-estimate $x(t)$, which is then used to find an updated polynomial function f to fit to P. This cycle is repeated until the fit to P ceases to improve significantly.

In the case of the Hammerstein model (Fig. 11.1), the fitting procedure begins with the deconvolution of P with $\dot{V}$ to yield an initial estimate of $h(t)$. The first estimate of the intermediate function $x(t)$ is obtained by deconvolving P with the estimated $h(t)$. A polynomial function of $\dot{V}$ is then fit to $x(t)$ to give the first estimate of f, after which the estimates of $h(t)$ and f are iteratively refined in a manner similar to that described above for the Wiener model.

Wiener and Hammerstein models have been fit to broad-band $\dot{V}$ and P data collected both from intact lungs [206, 209] and isolated strips of lung tissue [207]. Both models were found to provide very accurate descriptions of the nonlinearities in the data, most of which are due to the rheological properties of the lung tissue. The Wiener model seems to win out slightly in terms of overall accuracy, although the improvement over the Hammerstein model is rather minor [209]. In any case, the important point about the success of either model is that they show the static and dynamic mechanical properties of lung tissue to be separable. Furthermore, the nonlinear properties of lung tissue are, to a high degree of approximation, confined to its static behavior.

11.3   Lung tissue rheology

The separability of the linear dynamic behavior of lung tissue from its nonlinear static behavior is a curious phenomenon. Why should nature have chosen such a particular and highly constrained version of the Volterra series to account for the complex rheology of lung tissue, given the theoretically limitless range of possibilities inherent in Eq. 11.1? What is this telling us about the underlying physical processes taking place within the tissue as it undergoes bulk strain and shear? Indeed, one gets the sense that we may be looking at an important clue about how to link tissue structure to function, especially as separability of static and dynamic properties is not just seen in lung tissue. In fact,

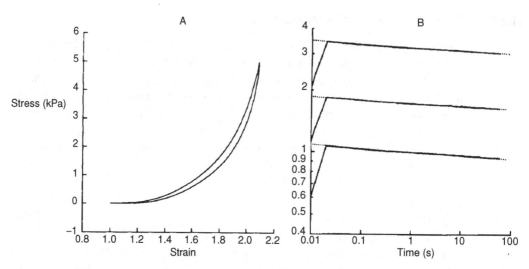

Figure 11.3 (A) Quasi-static stress-strain recording from a strip of lung tissue. (B) Stress-relaxation transients recorded in the same strip over the first 100 s following 10% step changes in strain from three different starting lengths (adapted from [181] with kind permission of Springer Science and Business Media).

a wide range of biological soft tissues exhibit this kind of behavior, including ligament [214] and heart valve tissues [215].

11.3.1   Quasi-linear viscoelasticity

The separability of static nonlinear behavior from dynamic linear behavior in biological soft tissue was first recognized by Fung [96], who coined the term *quasi-linear vis-coelasticity* to describe it. He noted that the stress response of strained tissue behaves as if the strain is first subjected to a zero-memory nonlinear transformation and then allowed to evolve according to a linear impulse response function – in other words, the Hammerstein model.

Quasi-linear viscoelasticity has become an accepted empirical model for many biological soft tissues. In the case of lung tissue, it is most easily appreciated in the step response. Figure 11.3A shows the quasi-static stress-strain curve obtained from a strip of degassed lung tissue subjected to a linear rate of change of strain to twice its unstressed length over a period of 50 s, and back again [181]. Apart from some slight looping due to incomplete resolution of viscoelastic transients, the stress-strain relationship is almost single-valued and is also highly nonlinear. In fact, the static stress-strain (σ-ε) relationship of the tissue strip is well described by the exponential function

$$\sigma(\varepsilon) = \alpha(e^{\beta\varepsilon} - 1) \tag{11.7}$$

where α and β are constants.

The step response of the linear dynamic system giving rise to stress adaptation in the same strips of lung tissue is apparent in Fig. 11.3B. Here we see that the logarithm of stress versus the logarithm of time describes an almost perfectly straight line. This means that the stress adaptation step response is described by t raised to some power k, a function we have already encountered in a similar context (Section 10.1.1). Thus,

$$\sigma(\varepsilon, t) = \alpha(e^{\beta\varepsilon} - 1)t^{-k}.$$
(11.8)

The question now is why does the dynamic mechanical behavior of lung tissue exhibit this intriguing functional form? Although the answer to this question is not known, it is instructive to delve a little more deeply into some of the considerations that may bear on it. This is the subject of the next section.

11.3.2 Power-law stress adaptation

We begin by addressing the question of why the stress in lung tissue following a step in strain should follow a power law in time. We have already seen that one possibility is to model power-law stress relaxation as arising from the ensemble behavior of numerous Maxwell bodies (Fig. 10.1). Such a system will exhibit power-law stress relaxation if the time-constants of the Maxwell bodies are distributed hyperbolically [96]. The problem is, this seems rather contrived. Why should we have to account for the genesis of power-law behavior in terms of one particular type of time-constant distribution? Maxwell body models like that shown in Fig. 10.1 also run into difficulties when they try to account for both power-law stress relaxation and static stress-strain nonlinearity at the same time. That is, although you can impose the appropriate time-constant distribution to generate power-law stress relaxation, as soon as you make the spring in a Maxwell body nonlinear it ceases to have a well-defined time-constant.

We encountered power-law behavior in Chapter 10 when discussing the constant phase model of lung impedance. The intact lung, with its air-liquid interface and complex three-dimensional structure, is certainly a very different physical system than a small strip of degassed parenchymal tissue. The appearance of power-law behavior in both systems thus speaks of the ubiquity of the phenomenon. In fact, nature abounds with processes that exhibit power-law dynamics. This is thought to reflect something fundamental about complex dynamic systems [216, 217]. Exactly what this might be remains something of a mystery, but several theories have been proposed. It has been suggested, for example, that what we often take to be a power-law distribution for the probability of some event may actually be the tail of a log-normal distribution, which rivals the Gaussian distribution in its ubiquity [218]. The log-normal distribution results when the stochastic event in question is the end result of a cascade of necessary prior stochastic events, a mechanism which certainly has the appropriate level of generality required of an explanation for power-law behavior. Another theory for the genesis of power-law behavior is self-organized criticality [217, 219], which holds that when a complex dynamic system receives a continual supply of some currency (be it energy, matter, stress, or whatever), the currency will tend to become internally organized near a critical state that poises the system on the brink of stability. Further supply will push the system beyond the stable limit, causing an avalanche of currency redistribution that re-establishes the near-critical state again [217]. Yet another source of power-law behavior that has recently received much attention is the "rich-get-richer" mechanism suggested to be behind the evolution of hubs in the internet [216, 220]. The idea here is that a small number of internet sites, for one reason or another, gain an unusual amount of early success, which feeds on itself to increase their notoriety in ever-increasing amounts. It has been shown that the rich-get-richer mechanism combined with probabilistic network growth can produce power-law behavior [216, 220, 221].

The common thread running though all the above theories for the genesis of power laws is that the appearance of the stochastic event in question can be attributed to a sequence of necessary antecedent events. Each antecedent event has its own probability of occurring given the chance, and must reach completion before the next event in line is initiated. This is not, however, a feature of conventional models of tissue stress adaptation in which collections of Maxwell bodies are arranged either in series or parallel. Consider, for example, the parallel collection of Maxwell bodies at the left-hand side of Fig. 11.4A, shown immediately after a step in strain. All the dashpots are fully compressed and all the springs are equally stretched. Subsequently, these Maxwell bodies relax at different rates so that at any later point in time their springs will be at various stages of decompression, such as in the situation shown at the right-hand side of Fig. 11.4A. Nevertheless, all Maxwell bodies continue to bear at least some of the total stress across the model at all times, rather than taking turns one at a time.

It is possible, however, to arrange Maxwell bodies so that stress is passed from one element to another in sequence [192], as shown in Fig. 11.4B. Here, we need to think of dashpots as models of physical pistons sliding inside cylinders, which can only elongate a finite amount before the piston slides out of the cylinder and the two components separate. Suppose a single Maxwell body initially bears the entire stress induced by a step increase in strain, and therefore also accounts for all the initial dynamics of stress relaxation (Fig. 11.4B, left-hand side). If the dashpot of this Maxwell body has a short length of travel, then its associated spring can only contract a small amount before the two moving parts of the dashpot disengage. To prevent the model from falling apart at this point, a second Maxwell body is connected between the piston of the first dashpot and the right-hand support of the model. Provided the length of the first dashpot is small enough, the second spring will have achieved only negligible extension by the time the first dashpot separates. Following separation, however, the spring of the second Maxwell body becomes extended so as to share the total extension equally with the first spring (Fig. 11.4B, right-hand side). However, because the extension of each spring is half the original extension of the first spring, the stress across the model is also commensurately lower. Stress relaxation continues to occur, but now it is the piston of the second dashpot that is doing the sliding, until it too reaches the end of its travel and disengages. A third Maxwell body then comes into play, reducing the common spring stress even further, and so on. In this way, the task of adapting to the stress across the model is passed in

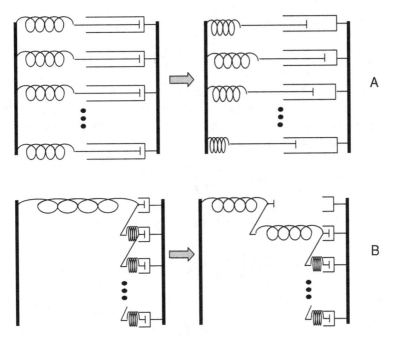

Figure 11.4 (A) A conventional viscoelastic model consisting of n Maxwell bodies arranged in parallel. Immediately following a step in strain, shown on the left, the dashpots (resistances R_1, $R_2, \ldots, R_n$) are all fully compressed while the springs (stiffnesses $E_1, E_2, \ldots, E_n$) are all equally extended. Some time later, shown on the right, the springs are at various stages of decompression according to their relative time-constants (the ratios R_i/E_i). (B) An alternative model of stress adaptation involving the sequential recruitment of Maxwell bodies. Immediately following a step in strain, the entire stress across the model is borne by a single spring, as shown on the left. After decompressing a short distance, the piston of the dashpot connected to this spring disengages from its cylinder, and a second spring is recruited to share the strain with the first spring, as shown on the right. This process continues as ever more springs are recruited to share the strain and transmit the stress. All springs have the same stiffness, E, and all dashpots have the same resistance, R (adapted from [192] with kind permission of Springer Science and Business Media).

turn from one Maxwell body to the next, each time with a sudden reduction in stress as a new spring is recruited to share the total extension.

The relaxation dynamics of the model in Fig. 11.4B are determined as follows [192]. Let the elastic behavior of each spring be given by

$$S = E \Delta x^\alpha \qquad (11.9)$$

where E is an elastic constant, Δx is the extension of the spring beyond its unstressed length, and α is a constant. When α is 1 the spring is Hookean, but arbitrary degrees of strain stiffening (typical of biological tissue) are permitted by increasing the value of α above unity. If the tissue consists of N identical springs connected in series, and it is stretched beyond its unstressed length by an amount L, then the extension of each individual spring is L/N and the common stress is

$$S = E \left(\frac{L}{N} \right)^{\alpha}. \tag{11.10}$$

When an additional spring becomes recruited into the line of stress-bearing elements by the mechanism illustrated in Fig. 11.4B, the total strain remains at L, but this is now shared equally by $N+1$ springs, each with extension $L/(N+1)$. Invoking the binomial theorem, we find the change in S produced by this recruitment to be

$$\Delta S = E \left(\frac{L}{N+1} \right)^{\alpha} - E \left(\frac{L}{N} \right)^{\alpha}$$

$$= E \left(\frac{L}{N} \right)^{\alpha} \left[\left(1 + \frac{1}{N} \right)^{-\alpha} - 1 \right]$$

$$= E \left(\frac{L}{N} \right)^{\alpha} \left[\frac{-\alpha}{N} + \frac{\alpha(\alpha+1)}{2!N^2} - \cdots \right]. \tag{11.11}$$

For large N this reduces to

$$\Delta S \approx -\frac{\alpha E}{N} \left(\frac{L}{N} \right)^{\alpha}$$

$$= -\frac{\alpha}{LE^{\frac{1}{\alpha}}} \left[E \left(\frac{L}{N} \right)^{\alpha} \right]^{\frac{\alpha+1}{\alpha}}$$

$$= -\frac{\alpha}{LE^{\frac{1}{\alpha}}} S^{1+\frac{1}{\alpha}}. \tag{11.12}$$

If the dashpots have a sufficiently short length of travel, D, then S remains essentially constant during the movement of each dashpot. The speed of movement, v, is inversely related to the dashpot resistance. So as not to be constrained by the requirement that this resistance be independent of velocity, we will let the resistive force generated within the dashpot be proportional to v raised to some power β. That is,

$$Rv^{\beta} = S. \tag{11.13}$$

But

$$v \approx \frac{D}{\Delta t} \tag{11.14}$$

where Δt is the time taken for the piston of the dashpot to move its full length of travel. Therefore,

$$\Delta t = DR^{\frac{1}{\beta}} S^{-\frac{1}{\beta}}. \tag{11.15}$$

Combining Eqs. 11.12 and 11.15 gives

$$\frac{\Delta S}{\Delta t} \rightarrow \frac{dS}{dt} = -\frac{\alpha}{LE^{\frac{1}{\alpha}} DR^{\frac{1}{\beta}}} S^{1+\frac{1}{\alpha}+\frac{1}{\beta}}, \tag{11.16}$$

the solution of which is

$$\int_{S_0}^{S} -\frac{LE^{\frac{1}{\alpha}}DR^{\frac{1}{\beta}}}{\alpha} S^{-\left(1+\frac{1}{\alpha}+\frac{1}{\beta}\right)} dS = \left[\frac{\beta LE^{\frac{1}{\alpha}}DR^{\frac{1}{\beta}}}{\alpha+\beta} S^{-\left(\frac{1}{\alpha}+\frac{1}{\beta}\right)}\right]_{S_0}^{S}$$

$$= \frac{\beta LE^{\frac{1}{\alpha}}DR^{\frac{1}{\beta}}}{\alpha+\beta} S^{-\left(\frac{1}{\alpha}+\frac{1}{\beta}\right)} - \frac{\beta LE^{\frac{1}{\alpha}}DR^{\frac{1}{\beta}}}{\alpha+\beta} S_0^{-\left(\frac{1}{\alpha}+\frac{1}{\beta}\right)}$$

$$= \frac{\beta LE^{\frac{1}{\alpha}}DR^{\frac{1}{\beta}}}{\alpha+\beta} S^{-\left(\frac{1}{\alpha}+\frac{1}{\beta}\right)} - \frac{\beta L^{-\frac{\alpha}{\beta}}E^{-\frac{1}{\beta}}DR^{\frac{1}{\beta}}}{\alpha+\beta} N_0^{\left(1+\frac{\alpha}{\beta}\right)}$$

$$= \int_0^t dt$$

$$= t \qquad (11.17)$$

where N_0 is the initial number of springs that bear the initial stress S_0. Rearranging gives

$$S = \left(\frac{\beta LE^{\frac{1}{\alpha}}DR^{\frac{1}{\beta}}}{\alpha+\beta}\right)^{\left(\frac{\alpha\beta}{\alpha+\beta}\right)} \left[\frac{\beta L^{-\frac{\alpha}{\beta}}E^{-\frac{1}{\beta}}DR^{\frac{1}{\beta}}N_0^{\left(1+\frac{\alpha}{\beta}\right)}}{\alpha+\beta} + t\right]^{-\left(\frac{\alpha\beta}{\alpha+\beta}\right)}. \qquad (11.18)$$

Equation 11.18 thus shows that the stress relaxation behavior of the model in Fig. 11.4B follows Eq. 11.8 as t becomes large, in which case

$$k = \frac{\alpha\beta}{\alpha+\beta}. \qquad (11.19)$$

Equation 11.18 will asymptote rapidly to a power law if D, the length of travel of each dashpot, is sufficiently small [192]. In other words, the amount of stress that is relaxed by the sliding of a dashpot is negligible compared to the amount of stress that is relaxed when the next spring is recruited.

Interestingly, this model also exhibits quasi-linear viscoelasticity because the exponent of the power-law relaxation (Eq. 11.19) depends only on the nonlinearities associated with the stiffness of the spring (α) and the resistance of the dashpot (β) in each Maxwell body, and is independent of strain. By incorporating this model into a larger construct involving sequential recruitment of collagen fibers, one can accurately predict experimental data [192].

It is certainly not clear at this point whether the above model has any basis in reality. Nevertheless, the model does serve to emphasize that the mechanical behavior of a complex material like lung tissue is likely to reflect the nature of the interactions between its components at least as much as the nature of the components themselves. Indeed, this characterizes other conceptions of the genesis of lung tissue mechanics, such as the model of nonlinear stress-strain behavior presented in Section 5.2.2 and models based on the theory of polymers [190]. It may be that all these various models have a role to play in the ultimate mechanistic description of lung tissue mechanics.

Problems

11.1 If $y(x) = Ax^2(t)$, what is $y(t)$ when $x(t) = \sin(\omega t)$? What does a system with this characteristic do to frequencies in the input? Can such a system be represented as either a Wiener or Hammerstein system? What is the harmonic distortion index? Is there any physiologically reasonable model of a lung that might behave in this way?

11.2 Using the methods outlined in Section 8.2.3, show that the second term in Eq. 11.1 transforms to the second term in Eq. 11.6.

11.3 Suppose that the forced oscillation technique is being used to track lung impedance continuously in a subject breathing at 12 breaths/min. The spontaneous breathing pattern has significant power up to the third harmonic. Lung impedance is to be measured above this frequency up to a maximum of 25 Hz. Specify a set of mutually prime frequencies for the forced oscillations that would cover this range with no more than 4 Hz between adjacent frequencies.

11.4 The power-law form of stress relaxation in the quasi-linear viscoelastic description of lung tissue following a step in strain (Eq. 11.8) means that lung tissue flows like a super-viscous fluid rather than behaving like an elastic solid. According to this description, if one waits long enough following a step in strain, the tissue stress should approach arbitrarily close to zero. This might seem to suggest that our lungs should eventually fall apart as a result of the continued stresses of breathing. From the time-courses of stress depicted in Fig. 11.3, estimate how long it would take for stress to reach 10% of its maximum value following stretch.

11.5 Equation 11.18 is a model for stress relaxation that exhibits quasi-linear behavior and is a power-law function of time for sufficiently large t. The value of k in Eq. 11.19 has been measured experimentally in strips of degassed lung tissue to be about 0.05. Also, lung tissue exhibits strain stiffening so that α has a value of 2 or more (Eq. 11.9). What does this say about the value of β? What is the nature of a resistive element described by Eq. 11.13 that would have such a value for β? Is there any physical system that behaves in this way?

12 Epilogue

This ends our tour through the inverse modeling world of lung mechanics, but it is only the tour that has ended. The inverse modeling world itself is limitless in extent, and its known borders continue to be extended by the synergistic forces of advancing experimental methods and increasing computational power. Nevertheless, the landscape we have covered in this book already exhibits widely varying terrain, and so it is easy to lose sight of the fact that all the different regions of this landscape are part of a single inverse modeling world. A brief overview will help bring it all together.

We began by pointing out in Chapter 1 that inverse models of lung mechanics can be usefully grouped into a hierarchy of increasing complexity (Fig. 1.6). The first level of complexity is represented by the single-compartment linear model, consisting of an elastic tissue compartment served by a single flow-resistive airway (Fig. 12.1). This model, which comprises the standard conception of lung mechanics, was explored in Chapter 3 and is characterized by the two parameters lung resistance (R_L) and lung elastance (E_L). Evaluation of R_L and E_L can be achieved by applying multiple linear regression to measurements of the independent variables volume (V) and flow ($\dot{V}$) together with the dependent variable transpulmonary pressure (P_{tp}). Confidence intervals about the estimated values of R_L and E_L can be determined according to classical regression theory under the assumptions that the noise in the data arises only in P_{tp} and that this noise is random, uncorrelated, and normally distributed.

Model fitting by multiple linear regression also assumes that the values of R_L and E_L are constant over the time during which data are collected. When this is not the case, one can resort to using the recursive form of multiple linear regression to track variations in R_L and E_L across the data set. The recursive approach can also be used to determine information-weighted histograms of R_L and E_L which characterize how these parameters vary when the single-compartment model has insufficient degrees of freedom to account for the data completely.

R_L is composed of two components, one being the flow resistance of the airways (R_{aw}) and the other reflecting viscous energy dissipation in the tissues (R_t). As discussed in Chapter 4, fluid dynamical theory provides a mechanistic link between the value of R_{aw} and the dimensions of an airway that is particularly well understood, especially at low Reynolds numbers when flow is laminar. The genesis of R_t is less well understood, but obviously somehow involves the dissipation of heat energy as the tissue is stretched and distorted.

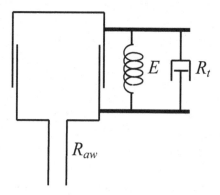

Figure 12.1 The single-compartment linear model. This model represents Complexity Level 1 in the world of inverse models of lung mechanics.

E_L arises in part through complex interactions between the many different constituents of lung tissue. However, the major determinant of E_L is the surface tension of the liquid that lines the inner surfaces of the lung. This follows as a consequence of Laplace's law. E_L is significantly reduced by the presence of pulmonary surfactant, is increased by derecruitment of lung volume, and is also affected by certain diseases of the parenchyma. R_L and E_L both increase during bronchoconstriction and so can be used as markers of the sensitivity and responsiveness of the lung to bronchial agonists.

We encounter a bifurcation in terms of how to extend the linear single-compartment model of the lung to the second level of complexity. One possibility, as discussed in Chapter 5, is to make the model nonlinear (Fig. 12.2A). For example, the resistance of the single airway could be made to depend on the flow through it. Alternatively, the elastance of the tissue compartment could be made to depend on the volume to which it is inflated. In either case, the result is an increase in the number of free parameters, which invariably results in a better fit of the model's predictions of P_{tp} to experimental measurements. This raises the critical question of how to decide when a more complicated model is a better model. The F-ratio test and the Akaike criterion are tools that can be brought to bear on this issue, and provide objective criteria for choosing between competing models. Even so, when any model of lung mechanics is fit to experiment measurements of V, $\dot{V}$, and P_{tp}, the residuals between the predicted and measured P_{tp} are frequently not random and uncorrelated, as required by classical regression theory. This raises theoretical difficulties for the calculation of confidence intervals about the estimated model parameter values. It also means that the F-ratio test and the Akaike criterion must be used cautiously as methods for discarding one model in favor of another.

Another key type of nonlinear mechanical behavior in the lung is the phenomenon of expiratory flow limitation. This phenomenon is critically important for diagnostic pulmonary medicine. Maximal expiratory flow depends on lung volume in a characteristic way in health, and is sensitive to the presence of obstructive and restrictive lung disease. As explained in Chapter 6, expiratory flow limitation can be accounted for both in terms of the dissipative mechanism of the choke point (Fig. 12.2B) and the related inertive mechanisms of the Bernoulli effect and wave speed.

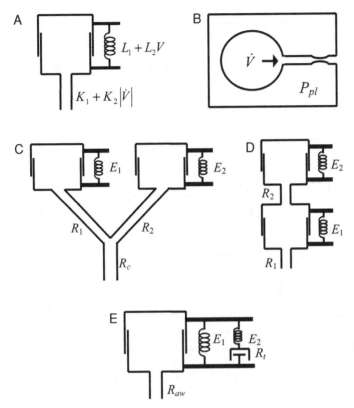

Figure 12.2 The models of lung mechanics representing Complexity Level 2. (A) The single-compartment nonlinear model. (B) A model of flow limitation by dynamic airway compression. (C) The parallel two-compartment linear model. (D) The series two-compartment linear model. (E) The linear viscoelastic model.

The second approach one can take to extending the single-compartment linear model is to keep the model linear while adding a second compartment. Here there are several physiologically plausible possibilities, as discussed in Chapter 7. For example, the two compartments can be connected either in parallel as in Fig. 12.2C, or in series as in Fig. 12.2D. It turns out, however, that the normal lung is better represented by a third possibility, the viscoelastic model shown in Fig. 12.2E. In this model, the second degree of mechanical freedom resides within the lung tissue which surrounds a single uniformly ventilated alveolar compartment. The differential equations describing these three models have exactly the same functional form, even though the models all have very different physiological interpretations. Consequently, these models cannot be distinguished on the basis of pressure-flow relationships measured at the airway opening. Resolving this model uniqueness problem required direct measurement of alveolar pressures.

Allowing a linear model of the lung to have an arbitrary number of compartments, instead of only two, takes us to the third level of inverse modeling complexity. Linear systems theory provides a set of mathematical tools for analyzing models of this type.

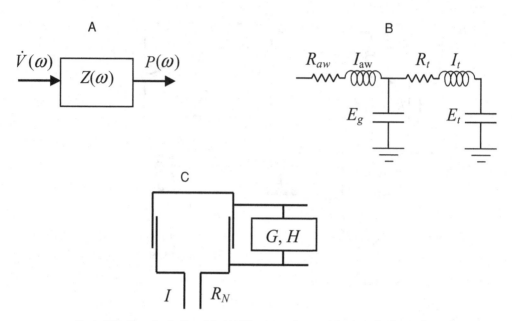

Figure 12.3 Complexity Level 3. (A) The general representation of a linear dynamic system in which impedance (Z) as a function of angular frequency (ω) links an input flow to an output pressure. (B) An electrical circuit representation of the six-element model of the lung with parameters characterizing the separate properties of airways and tissues. (C) The constant phase model of the lung featuring a single airway serving a homogeneously ventilated alveolar compartment composed of tissue with impedance characterized by real part G and imaginary part H.

Because of the principle of superposition, a linear dynamic system is completely characterized by its frequency response, which specifies how input sine waves are altered in amplitude and phase to produce the output. The impulse response, and its integral the step response, also completely characterize a linear dynamic system by allowing the output to be calculated from an arbitrary input via the convolution integral. The most important tools in linear systems theory are the Fourier transform and its numerical counterpart the FFT. The Fourier transform of the impulse response is the frequency response, also called a transfer function. When the input is flow and the output is pressure, the transfer function of a linear dynamic system is called mechanical impedance. This is the conceptual basis for characterizing lung mechanics in terms of impedance (Fig. 12.3A). The forced oscillation technique is the general experimental approach of applying broad-band perturbations in flow to the lungs so that impedance can be determined at multiple frequencies simultaneously. This applies to both input impedance and transfer impedance.

The differential equations of motion of compartment models of the lung can also be subjected to Fourier transformation. The input impedance of the single-compartment model has a real part, called resistance, that is independent of frequency. The imaginary part of its impedance, called reactance, crosses the frequency axis at the resonant frequency. As well as being applied to the entire lung, the single-compartment model can

$$y(t) = h_0 + \int_0^\infty h_1(u_1)x(t-u_1)du_1 + \int_0^\infty \int_0^\infty h_2(u_1,u_2)x(t-u_1)x(t-u_2)du_1du_2$$

$$+\int_0^\infty \int_0^\infty \int_0^\infty h_3(u_1,u_2,u_3)x(t-u_1)x(t-u_2)x(t-u_3)du_1du_2du_3 + \cdots$$

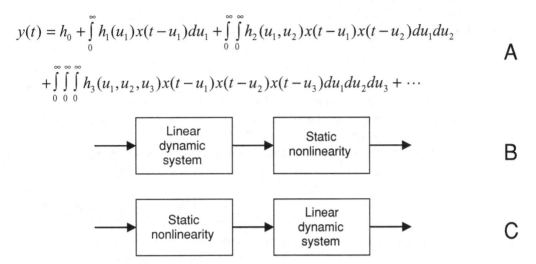

Figure 12.4 Complexity Level 4. (A) The Volterra series. (B) The Wiener model. (C) The Hammerstein model.

also be used to describe the impedance of a tiny sub-pleural region of the lung accessed through an alveolar capsule.

Measurements of the low-frequency input impedance of the lung show that resistance usually decreases slightly as frequency increases, especially in disease. This imposes the need to invoke additional compartments in the model. Most of the frequency dependence of resistance above 4 Hz appears to be due to heterogeneities of regional ventilation. Below 2 Hz, the marked negative dependence of resistance on frequency can be ascribed largely to the viscoelastic properties of the lung tissue, at least in a normal lung. The six-element model of the lung (Fig. 12.3B) has been used to account for the main features of both input and transfer impedances above 4 Hz until frequencies reach those at which acoustic phenomena in the airways have to be taken into account.

The constant phase model of lung tissue impedance, and its pulmonary extension that includes airway impedance (Fig. 12.3C), are other linear dynamic models of lung mechanics. The constant phase model arose from observations that stress adaptation in tissue is much more accurately described by a power law than by a decaying exponential function of time. The constant phase model has been very successful for describing the input impedance of the lung below 20 Hz, and provides a characterization of lung mechanics in terms of only four parameters: R_N and I characterize the airways, while G and H characterize the tissues. The constant phase model has been extended to include continuous distributions of airway resistances or tissue elastances in parallel to describe impedance when the lung becomes heterogeneous. The presence of heterogeneities can also often be confidently inferred from an increase in hysteresivity, defined as the ratio G/H. The constant phase model, however, does not fit easily into the formalism of ordinary differential equations. Instead, it invokes an equation involving a fractional derivative of time. Applications of the constant phase model are numerous in the study of pulmonary pathophysiology, conveniently allowing lung mechanics to be partitioned between the conducting airways and the lung periphery.

With nonlinear dynamic models, one enters the final frontier in the inverse modeling world. Even the Volterra series (Fig. 12.4A) cannot model every conceivable nonlinear dynamic system because it does not account for infinite memory. Within that constraint, however, it is still not practical to model lung mechanics with an infinity of terms of ever-increasing complexity. Block-structured subsets of the Volterra series known as the Wiener and Hammerstein models (Figs. 12.4B and C, respectively) have been used to model lung mechanics with significant precision. These models show that lung tissue exhibits the peculiar property known as quasi-linear viscoelasticity. The mechanistic basis of this behavior is unclear.

Finally, if there is one overall message that should be taken away from reading this book, it is that there is no definitively correct inverse model of lung mechanics. Indeed, the very concept of model correctness is fundamentally flawed. Models merely represent our current understanding of how a system behaves. Furthermore, how a system behaves depends on the conditions under which it is examined. We have seen, for example, that increasing the frequency range of lung impedance measurements increases the reliability with which the parameters of a model can be estimated. We have also seen, however, that new physical phenomena come into play when the range of frequencies is increased. Accounting for these new phenomena means incorporating additional structure into the model. This, in turn, increases the number of free parameters that must be evaluated. Our understanding of lung mechanics thus progresses through a never-ending competition between the search for novel ways of testing models and the development of novel models for describing the data that are collected.

References

1. Jackson, A.C. and A. Vinegar, A technique for measuring frequency response of pressure, volume, and flow transducers. *J Appl Physiol*, 1979. **47**: p. 462–467.
2. Duvivier, C., et al., Static and dynamic performances of variable reluctance and piezoresistive pressure transducers for forced oscillation measurements. *Eur Respir J*, 1991. **1**: p. 146–150.
3. Pedley, T. and J. Drazen, Aerodynamic theory. In *Handbook of Physiology. Section 3: The Respiratory System*. P. Macklem and J. Mead, Editors. 1986, Bethesda, MD: American Physiological Society, p. 41–54.
4. Bates, J.H.T., et al., Correcting for the Bernoulli effect in lateral pressure measurements. *Pediatr Pulmonol*, 1992. **12**: p. 251–256.
5. Bates, J.H. and A.M. Lauzon, A nonstatistical approach to estimating confidence intervals about model parameters: application to respiratory mechanics. *IEEE Trans Biomed Eng*, 1992. **39**(1): p. 94–100.
6. Navalesi, P., et al., Influence of site of tracheal pressure measurement on in situ estimation of endotracheal tube resistance. *J Appl Physiol*, 1994. **77**: p. 2899–2906.
7. Baydur, A., et al., A simple method for assessing the validity of the esophageal balloon technique. *Am Rev Respir Dis*, 1982. **126**: p. 788–791.
8. Dechman, G., J. Sato, and J.H.T. Bates, Factors affecting the accuracy of esophageal balloon measurement of pleural pressure in dogs. *J Appl Physiol*, 1992. **72**: p. 383–388.
9. Peslin, R., et al., Validity of the esophageal balloon technique at high frequencies. *J Appl Physiol*, 1993. **74**: p. 1039–1044.
10. Panizza, J.A., Comparison of balloon and transducer catheters for estimating lung elasticity. *J Appl Physiol*, 1992. **72**: p. 231–235.
11. Wang, C.G., et al., Methacholine-induced airway reactivity of inbred rats. *J Appl Physiol*, 1986. **61**(6): p. 2180–2185.
12. Fredberg, J.J., et al., Alveolar pressure nonhomogeneity during small-amplitude high-frequency oscillation. *J Appl Physiol*, 1984. **57**(3): p. 788–800.
13. Bates, J.H., P. Baconnier, and J. Milic-Emili, A theoretical analysis of interrupter technique for measuring respiratory mechanics. *J Appl Physiol*, 1988. **64**(5): p. 2204–2214.
14. Ludwig, M.S., et al., Partitioning of pulmonary resistance during constriction in the dog: effects of volume history. *J Appl Physiol*, 1987. **62**(2): p. 807–815.
15. Ludwig, M.S., P.V. Romero, and J.H. Bates, A comparison of the dose-response behavior of canine airways and parenchyma. *J Appl Physiol*, 1989. **67**(3): p. 1220–1225.
16. Tomioka, S., J.H. Bates, and C.G. Irvin, Airway and tissue mechanics in a murine model of asthma: alveolar capsule vs. forced oscillations. *J Appl Physiol*, 2002. **93**(1): p. 263–270.
17. Bates, J.H., et al., Measurement of alveolar pressure in closed-chest dogs during flow interruption. *J Appl Physiol*, 1989. **67**(1): p. 488–492.

18. Renzi, P.E., C.A. Giurdanella, and A.C. Jackson, Improved frequency response of pneumo-tachometers by digital compensation. *J Appl Physiol*, 1990. **68**: p. 382–386.
19. Farre, R., et al., Analysis of the dynamic characteristics of pressure transducers for studying respiratory mechanics at high frequencies. *Med & Biol Eng & Comput*, 1989. **27**: p. 531–536.
20. Schuessler, T.F., G.N. Maksym, and J.H.T. Bates, Estimating tracheal flow in small animals. *Proceedings of the 15th Annual International Meeting of the IEEE Engineering in Medicine and Biology Society*, San Diego, October 28–31, 1993: p. 560–561.
21. Schibler, A., et al., Measurement of lung volume and ventilation distribution with an ultra-sonic flow meter in healthy infants. *Eur Respir J*, 2002. **20**(4): p. 912–918.
22. Clary, A.L. and J.M. Fouke, Fast-responding automated airway temperature probe. *Med & Biol Eng & Comput*, 1991. **29**: p. 501–504.
23. Cohen, K.P., et al., Comparison of impedance and inductance ventilation sensors on adults during breathing, motion, and simulated airway obstruction. *IEEE Trans Biomed Eng*, 1997. **44**: p. 555–566.
24. Aliverti, A., et al., Compartmental analysis of breathing in the supine and prone positions by optoelectronic plethysmography. *Ann Biomed Eng*, 2001. **29**: p. 60–70.
25. Bates, J.H.T., Measurement techniques in respiratory mechanics. In *Physiologic Basis of Respiratory Disease*. Q. Hamid, J. Shannon, and J. Martin, Editors. 2005, Hamilton, Canada: BC Decker, p. 623–637.
26. Lundblad, L.K., et al., Airway hyperresponsiveness in allergically inflamed mice: the role of airway closure. *Am J Respir Crit Care Med*, 2007. **175**(8): p. 768–774.
27. Gold, W., Pulmonary function testing. In *Textbook of Respiratory Medicine*. J. Murray and J. Nadel, Editors. 2000, Philadelphia, PA: Saunders, p. 793–799.
28. Shore, S., J. Milic-Emili, and J.G. Martin, Reassessment of body plethysmographic technique for the measurement of thoracic gas volume in asthmatics. *Am Rev Respir Dis*, 1982. **126**(3): p. 515–520.
29. Dechman, G.S., et al., The effect of changing end-expiratory pressure on respiratory system mechanics in open- and closed-chest anesthetized, paralyzed patients. *Anesth Analg*, 1995. **81**(2): p. 279–286.
30. Marini, J.J., Auto-positive end-expiratory pressure and flow limitation in adult respiratory distress syndrome–intrinsically different? *Crit Care Med*, 2002. **30**(9): p. 2140–2141.
31. Von Neergaard, K. and K. Wirz, Die Messung der Stromungswiderstande in den Atemwegen des Menschen, insbesondere bei Asthma und Emphysem. *Ztschr f Klin Med*, 1927. **105**: p. 51–82.
32. Mols, G., et al., Volume-dependent compliance in ARDS: proposal of a new diagnostic concept. *Intensive Care Med*, 1999. **25**(10): p. 1084–1091.
33. Mead, J. and J.L. Whittenberger, Evaluation of airway interruption technique as a method for measuring pulmonary airflow resistance. *J Appl Physiol*, 1954. **6**(7): p. 408–416.
34. Frank, N.R., J. Mead, and J.L. Whittenberger, Comparative sensitivity of four methods for measuring changes in respiratory flow resistance in man. *J Appl Physiol*, 1971. **31**(6): p. 934–938.
35. Lauzon, A.M. and J.H. Bates, Estimation of time-varying respiratory mechanical parameters by recursive least squares. *J Appl Physiol*, 1991. **71**(3): p. 1159–1165.
36. Lee, R.C.K., *Optimal Estimation, Identification and Control*. 1964, Cambridge, MA: MIT Press.
37. Hantos, Z., et al., Parameter estimation of transpulmonary mechanics by a nonlinear inertive model. *J Appl Physiol*, 1982. **52**(4): p. 955–963.

38. Bijaoui, E.L., et al., Mechanical properties of the lung and upper airways in patients with sleep-disordered breathing. *Am J Respir Crit Care Med*, 2002. **165**(8): p. 1055–1061.

39. West, J., *Respiratory Physiology. The Essentials*. 6th edition. 1999, Philadelphia, PA: Lippincott Williams & Wilkins.

40. Leff, A. and P. Schumacker, *Respiratory Physiology. Basics and Applications*. 1993, Philadelphia, PA: Saunders.

41. Pedley, T.J., R.C. Schroter, and M.F. Sudlow, The prediction of pressure drop and variation of resistance within the human bronchial airways. *Respir Physiol*, 1970. **9**(3): p. 387–405.

42. Kabilan, S., C.L. Lin, and E.A. Hoffman, Characteristics of airflow in a CT-based ovine lung: a numerical study. *J Appl Physiol*, 2007. **102**(4): p. 1469–1482.

43. Weibel, E., *Morphometry of the Human Lung*. 1963, Berlin: Springer.

44. Horsfield, K., W. Kemp, and S. Phillips, An asymmetrical model of the airways of the dog lung. *J Appl Physiol*, 1982. **52**(1): p. 21–26.

45. Fredberg, J. and A. Hoenig, Mechanical response of the lung at high frequencies. *J Biomech Eng*, 1978. **100**: p. 57–66.

46. Lutchen, K.R., J.L. Greenstein, and B. Suki, How inhomogeneities and airway walls affect frequency dependence and separation of airway and tissue properties. *J Appl Physiol*, 1996. **80**(5): p. 1696–1707.

47. Thorpe, C.W. and J.H. Bates, Effect of stochastic heterogeneity on lung impedance during acute bronchoconstriction: a model analysis. *J Appl Physiol*, 1997. **82**(5): p. 1616–1625.

48. Gillis, H.L. and K.R. Lutchen, How heterogeneous bronchoconstriction affects ventilation distribution in human lungs: a morphometric model. *Ann Biomed Eng*, 1999. **27**(1): p. 14–22.

49. Gomes, R.F. and J.H. Bates, Geometric determinants of airway resistance in two isomorphic rodent species. *Respir Physiol Neurobiol*, 2002. **130**(3): p. 317–325.

50. Kaczka, D.W., et al., Airway and lung tissue mechanics in asthma. Effects of albuterol. *Am J Respir Crit Care Med*, 1999. **159**(1): p. 169–178.

51. Ebihara, T., et al., Changes in extracellular matrix and tissue viscoelasticity in bleomycin-induced lung fibrosis. Temporal aspects. *Am J Respir Crit Care Med*, 2000. **162**(4 Pt 1): p. 1569–1576.

52. Bates, J.H., A. Cojocaru, and L.K. Lundblad, Bronchodilatory effect of deep inspiration on the dynamics of bronchoconstriction in mice. *J Appl Physiol*, 2007. **103**(5): p. 1696–1705.

53. Lauzon, A.M. and J.H. Bates, Kinetics of respiratory system elastance after airway challenge in dogs. *J Appl Physiol*, 2000. **89**(5): p. 2023–2029.

54. Bates, J.H., et al., Temporal dynamics of pulmonary response to intravenous histamine in dogs: effects of dose and lung volume. *J Appl Physiol*, 1994. **76**(2): p. 616–626.

55. Mitzner, W., et al., Effect of bronchial smooth muscle contraction on lung compliance. *J Appl Physiol*, 1992. **72**(1): p. 158–167.

56. Bates, J.H., F.A. Donoso, and R. Peslin, Airway and tissue impedances of canine lungs after step volume changes. *J Appl Physiol*, 1993. **75**(4): p. 1460–1466.

57. Kelly, S.M., J.H. Bates, and R.P. Michel, Altered mechanical properties of lung parenchyma in postobstructive pulmonary vasculopathy. *J Appl Physiol*, 1994. **77**(6): p. 2543–2551.

58. Verbeken, E.K., et al., Structure and function in fibrosing alveolitis. *J Appl Physiol*, 1994. **76**(2): p. 731–742.

59. Takubo, Y., et al., Alpha1-antitrypsin determines the pattern of emphysema and function in tobacco smoke-exposed mice: parallels with human disease. *Am J Respir Crit Care Med*, 2002. **166**(12 Pt 1): p. 1596–1603.

60. Verbeken, E.K., et al., Tissue and airway impedance of excised normal, senile, and emphysematous lungs. *J Appl Physiol*, 1992. **72**(6): p. 2343–2353.

61. Hubmayr, R.D., Perspective on lung injury and recruitment: a skeptical look at the opening and collapse story. *Am J Respir Crit Care Med*, 2002. **165**(12): p. 1647–1653.

62. Halpern, D. and J.B. Grotberg, Surfactant effects on fluid-elastic instabilities of liquid-lined flexible tubes: a model of airway closure. *J Biomech Eng*, 1993. **115**(3): p. 271–277.

63. Allen, G.B., et al., Pulmonary impedance and alveolar instability during injurious ventilation in rats. *J Appl Physiol*, 2005. **99**(2): p. 723–730.

64. Crotti, S., et al., Recruitment and derecruitment during acute respiratory failure. An experimental study. *Am J Respir Cell Mol Biol*, 2001. **164**: p. 131–140.

65. Gattinoni, L., et al., Effects of positive end-expiratory pressure on regional distribution of tidal volume and recruitment in adult respiratory distress syndrome. *Am J Respir Crit Care Med*, 1995. **151**(6): p. 1807–1814.

66. King, G.G., et al., Differences in airway closure between normal and asthmatic subjects measured with single-photon emission computed tomography and technegas. *Am J Respir Crit Care Med*, 1998. **158**(6): p. 1900–1906.

67. Christie, R.V., The elastic properties of the emphysematous lung and their clinical significance. *J Clin Invest*, 1934. **13**(2): p. 295–321.

68. Bachofen, H., J. Hildebrandt, and M. Bachofen, Pressure-volume curves of air- and liquid-filled excised lungs-surface tension in situ. *J Appl Physiol*, 1970. **29**(4): p. 422–431.

69. Prange, H.D., Laplace's law and the alveolus: a misconception of anatomy and a misapplication of physics. *Adv Physiol Educ*, 2003. **27**(1–4): p. 34–40.

70. Allen, G., et al., Transient mechanical benefits of a deep inflation in the injured mouse lung. *J Appl Physiol*, 2002. **93**(5): p. 1709–1715.

71. Lauzon, A.M., G. Dechman, and J.H. Bates, Time course of respiratory mechanics during histamine challenge in the dog. *J Appl Physiol*, 1992. **73**(6): p. 2643–2647.

72. Ingram, R.H. and T.J. Pedley, Pressure-flow relationships in the lungs. In *Handbook of Physiology. Section 3: The Respiratory System*. P. Macklem and J. Mead, Editors. 1986, Bethesda, MD: American Physiological Society, p. 277–293.

73. Lutchen, K.R., et al., Airway inhomogeneities contribute to apparent lung tissue mechanics during constriction. *J Appl Physiol*, 1996. **80**(5): p. 1841–1849.

74. Mishima, M., Z. Balassy, and J.H. Bates, Acute pulmonary response to intravenous histamine using forced oscillations through alveolar capsules in dogs. *J Appl Physiol*, 1994. **77**(5): p. 2140–2148.

75. Ding, D.J., J.G. Martin, and P.T. Macklem, Effects of lung volume on maximal methacholine-induced bronchoconstriction in normal humans. *J Appl Physiol*, 1987. **62**(3): p. 1324–1330.

76. An, S.S., et al., Airway smooth muscle dynamics: a common pathway of airway obstruction in asthma. *Eur Respir J*, 2007. **29**(5): p. 834–860.

77. Brusasco, V. and R. Pellegrino, Complexity of factors modulating airway narrowing in vivo: relevance to assessment of airway hyperresponsiveness. *J Appl Physiol*, 2003. **95**(3): p. 1305–1313.

78. McParland, B.E., P.T. Macklem, and P.D. Pare, Airway wall remodeling: friend or foe? *J Appl Physiol*, 2003. **95**(1): p. 426–434.

79. Bates, J.H. and A.M. Lauzon, Parenchymal tethering, airway wall stiffness, and the dynamics of bronchoconstriction. *J Appl Physiol*, 2007. **102**(5): p. 1912–1920.

80. Moreno, R.H., J.C. Hogg, and P.D. Pare, Mechanics of airway narrowing. *Am Rev Respir Dis*, 1986. **133**(6): p. 1171–1180.

81. Yager, D., et al., Amplification of airway constriction due to liquid filling of airway interstices. *J Appl Physiol*, 1989. **66**(6): p. 2873–2884.

82. Wiggs, B.R., et al., A model of airway narrowing in asthma and in chronic obstructive pulmonary disease. *Am Rev Respir Dis*, 1992. **145**(6): p. 1251–1258.

83. Bates, J.H., et al., The synergistic interactions of allergic lung inflammation and intratracheal cationic protein. *Am J Respir Crit Care Med*, 2008. **177**(3): p. 261–268.

84. Brown, K., et al., Evaluation of the flow-volume loop as an intra-operative monitor of respiratory mechanics in infants. *Pediatr Pulmonol*, 1989. **6**(1): p. 8–13.

85. Wagers, S., et al., Nonlinearity of respiratory mechanics during bronchoconstriction in mice with airway inflammation. *J Appl Physiol*, 2002. **92**(5): p. 1802–1807.

86. Bersten, A.D., Measurement of overinflation by multiple linear regression analysis in patients with acute lung injury. *Eur Respir J*, 1998. **12**: p. 526–532.

87. Salazar, E. and J.H. Knowles, An analysis of pressure-volume characteristics of the lungs. *J Appl Physiol*, 1964. **19**: p. 97–104.

88. Eidelman, D.H., H. Ghezzo, and J.H. Bates, Exponential fitting of pressure-volume curves: confidence limits and sensitivity to noise. *J Appl Physiol*, 1990. **69**(4): p. 1538–1541.

89. Venegas, J.G., R.S. Harris, and B.A. Simon, A comprehensive equation for the pulmonary pressure-volume curve. *J Appl Physiol*, 1998. **84**(1): p. 389–395.

90. Goerke, J. and J.A. Clements, Alveolar surface tension and lung surfactant. In *Handbook of Physiology. Section 3: The Respiratory System*. P. Macklem and J. Mead, Editors. 1986, Bethesda, MD: American Physiological Society, p. 247–261.

91. Otis, D.R., Jr., et al., Role of pulmonary surfactant in airway closure: a computational study. *J Appl Physiol*, 1993. **75**(3): p. 1323–1333.

92. Allen, G. and J.H. Bates, Dynamic mechanical consequences of deep inflation in mice depend on type and degree of lung injury. *J Appl Physiol*, 2004. **96**(1): p. 293–300.

93. Mead, J., Mechanical properties of lungs. *Physiol Rev*, 1961. **41**: p. 281–330.

94. Fung, Y.C., Microrheology and constitutive equation of soft tissue. *Biorheology*, 1988. **25**(1–2): p. 261–270.

95. Haut, R.C. and R.W. Little, A constitutive equation for collagen fibers. *J Biomech*, 1972. **5**(5): p. 423–430.

96. Fung, Y., *Biomechanics. Mechanical Properties of Living Tissues*. 1981, New York, NY: Springer-Verlag, p. 41–53.

97. Sobin, S.S., Y.C. Fung, and H.M. Tremer, Collagen and elastin fibers in human pulmonary alveolar walls. *J Appl Physiol*, 1988. **64**(4): p. 1659–1675.

98. Maksym, G.N. and J.H. Bates, A distributed nonlinear model of lung tissue elasticity. *J Appl Physiol*, 1997. **82**(1): p. 32–41.

99. Suki, B. and J.H. Bates, Extracellular matrix mechanics in lung parenchymal diseases. *Respir Physiol Neurobiol*, 2008. **163**(1–3): p. 33–43.

100. Maksym, G.N., J.J. Fredberg, and J.H. Bates, Force heterogeneity in a two-dimensional network model of lung tissue elasticity. *J Appl Physiol*, 1998. **85**(4): p. 1223–1229.

101. Bates, J.H., et al., Linking parenchymal disease progression to changes in lung mechanical function by percolation. *Am J Respir Crit Care Med*, 2007. **176**(6): p. 617–623.

102. Collard, H.R., et al., Acute exacerbations of idiopathic pulmonary fibrosis. *Am J Respir Crit Care Med*, 2007. **176**(7): p. 636–643.

103. Pagano, M. and K. Gauvreau, Analysis of variance. In *Principles of Biostatistics*. 2000, Pacific Grove, CA: Duxbury, p. 288–290.

104. Akaike, H., A new look at the statistical model identification. *IEEE Trans Automatic Control*, 1974. **16**(6): p. 716–723.

105. Hurvich, C.M. and C.-L. Tsai, Regression and time series model selection in small samples. *Biometrika*, 1989. **76**: p. 297–307.

106. Kaczka, D.W., C.B. Massa, and B.A. Simon, Reliability of estimating stochastic lung tissue heterogeneity from pulmonary impedance spectra: a forward-inverse modeling study. *Ann Biomed Eng*, 2007. **35**(10): p. 1722–1738.

107. Hyatt, R., Forced expiration. In *Handbook of Physiology. Section 3: The Respiratory System*. P. Macklem and J. Mead, Editors. 1986, Bethesda, MD: American Physiological Society, p. 295–314.

108. Fish, J., Bronchial challenge testing. In *Allergy: Principles and Practice*. 4th edition. E. Middleton, Editor. 1993, St. Louis, MO: Mosby-Year Book.

109. Wilson, T., J. Rodarte, and J. Butler, Wave-speed and viscous flow limitation. In *Handbook of Physiology. Section 3: The Respiratory System*. P. Macklem and J. Mead, Editors. 1986, Bethesda, MD: American Physiological Society, p. 295–314.

110. Bates, J.H.T., Physics of expiratory flow limitation. In *Physiologic Basis of Respiratory Disease*. Q. Hamid, J. Shannon, and J. Martin, Editors. 2005, Hamilton, Canada: BC Decker, p. 55–60.

111. Du Toit, J.I., et al., Characteristics of bronchial hyperresponsiveness in smokers with chronic air-flow limitation. *Am Rev Respir Dis*, 1986. **134**(3): p. 498–501.

112. Cherniak, R., *Pulmonary Function Testing*. 1977, Phildelphia, PA: WB Saunders.

113. Polak, A.G. and K.R. Lutchen, Computational model for forced expiration from asymmetric normal lungs. *Ann Biomed Eng*, 2003. **31**(8): p. 891–907.

114. Lambert, R.K., Simulation of the effects of mechanical nonhomogeneities on expiratory flow from human lungs. *J Appl Physiol*, 1990. **68**(6): p. 2550–2563.

115. Macklem, P.T. and J. Mead, Factors determining maximum expiratory flow in dogs. *J Appl Physiol*, 1968. **25**(2): p. 159–169.

116. Dawson, S.V. and E.A. Elliott, Wave-speed limitation on expiratory flow – a unifying concept. *J Appl Physiol*, 1977. **43**(3): p. 498–515.

117. Mink, S., M. Ziesmann, and L.D. Wood, Mechanisms of increased maximum expiratory flow during HeO_2 breathing in dogs. *J Appl Physiol*, 1979. **47**(3): p. 490–502.

118. Weinberger, H., *A First Course in Partial Differential Equations*. 1965, New York, NY: Wiley.

119. Pride, N.B., et al., Determinants of maximal expiratory flow from the lungs. *J Appl Physiol*, 1967. **23**(5): p. 646–662.

120. Zamir, M., Pulsatile flow in an elastic tube. In *The Physics of Pulsatile Flow*. 2000, New York, NY: Springer-Verlag, p. 113–146.

121. Foss, S.D., A method of exponential curve fitting by numerical integration. *Biometrics*, 1970. **26**: p. 815–821.

122. Bates, J.H., et al., Respiratory resistance with histamine challenge by single-breath and forced oscillation methods. *J Appl Physiol*, 1986. **61**(3): p. 873–880.

123. Bates, J.H., et al., Volume-time profile during relaxed expiration in the normal dog. *J Appl Physiol*, 1985. **59**(3): p. 732–737.

124. Otis, A.B., et al., Mechanical factors in distribution of pulmonary ventilation. *J Appl Physiol*, 1956. **8**(4): p. 427–443.

125. Mead, J., Contribution of compliance of airways to frequency-dependent behavior of lungs. *J Appl Physiol*, 1969. **26**(5): p. 670–673.

126. Macklem, P.T., Airway obstruction and collateral ventilation. *Physiol Rev*, 1971. **51**(2): p. 368–436.

127. Mount, L.E., The ventilation flow-resistance and compliance of rat lungs. *J Physiol*, 1955. **127**(1): p. 157–167.

128. Bates, J.H., A. Rossi, and J. Milic-Emili, Analysis of the behavior of the respiratory system with constant inspiratory flow. *J Appl Physiol*, 1985. **58**(6): p. 1840–1848.

129. Similowski, T. and J.H. Bates, Two-compartment modelling of respiratory system mechanics at low frequencies: gas redistribution or tissue rheology? *Eur Respir J*, 1991. **4**(3): p. 353–358.

130. D'Angelo, E., et al., Respiratory mechanics in anesthetized paralyzed humans: effects of flow, volume, and time. *J Appl Physiol*, 1989. **67**(6): p. 2556–2564.

131. Bates, J.H., et al., Interrupter resistance elucidated by alveolar pressure measurement in open-chest normal dogs. *J Appl Physiol*, 1988. **65**(1): p. 408–414.

132. Romero, P.V., et al., High-frequency characteristics of respiratory mechanics determined by flow interruption. *J Appl Physiol*, 1990. **69**(5): p. 1682–1688.

133. Fredberg, J.J., et al., Nonhomogeneity of lung response to inhaled histamine assessed with alveolar capsules. *J Appl Physiol*, 1985. **58**(6): p. 1914–1922.

134. Ludwig, M.S., et al., Interpretation of interrupter resistance after histamine-induced constriction in the dog. *J Appl Physiol*, 1990. **68**(4): p. 1651–1656.

135. Sato, J., et al., Low-frequency respiratory system resistance in the normal dog during mechanical ventilation. *J Appl Physiol*, 1991. **70**(4): p. 1536–1543.

136. D'Angelo, E., M. Tavola, and J. Milic-Emili, Volume and time dependence of respiratory system mechanics in normal anaesthetized paralysed humans. *Eur Respir J*, 2000. **16**(4): p. 665–672.

137. Cooley, J.W. and J.W. Tukey, An algorithm for the machine calculation of complex Fourier series. *Math Comput*, 1965. **19**: p. 297–301.

138. Dubois, A.B., et al., Oscillation mechanics of lungs and chest in man. *J Appl Physiol*, 1956. **8**(6): p. 587–594.

139. Peslin, R. and J. Fredberg, Oscillation mechanics of the respiratory system. In *Handbook of Physiology. Section 3: The Respiratory System*. P. Macklem and J. Mead, Editors. 1986, Bethesda, MD: American Physiological Society, p. 145–178.

140. Zwart, A. and K.P. Vaqn de Woestijne, Mechanical respiratory impedance by forced oscillation. *European Respiratory Review*, 1994. **4**: p. 114–237.

141. Ferris, B.G., Jr., J. Mead, and L.H. Opie, Partitioning of respiratory flow resistance in man. *J Appl Physiol*, 1964. **19**: p. 653–658.

142. Franken, H., et al., Forced oscillation technique: comparison of two devices. *J Appl Physiol*, 1985. **59**(5): p. 1654–1659.

143. Peslin, R., et al., Respiratory impedance measured with head generator to minimize upper airway shunt. *J Appl Physiol*, 1985. **59**(6): p. 1790–1795.

144. Lutchen, K.R., et al., Use of transfer impedance measurements for clinical assessment of lung mechanics. *Am J Respir Crit Care Med*, 1998. **157**(2): p. 435–446.

145. Kaminsky, D.A., et al., Oscillation mechanics of the human lung periphery in asthma. *J Appl Physiol*, 2004. **97**(5): p. 1849–1858.

146. Davey, B.L. and J.H. Bates, Regional lung impedance from forced oscillations through alveolar capsules. *Respir Physiol*, 1993. **91**(2–3): p. 165–182.

147. Bijaoui, E., P.F. Baconnier, and J.H. Bates, Mechanical output impedance of the lung determined from cardiogenic oscillations. *J Appl Physiol*, 2001. **91**(2): p. 859–865.

148. Schuessler, T.F. and J.H. Bates, A computer-controlled research ventilator for small animals: design and evaluation. *IEEE Trans Biomed Eng*, 1995. **42**(9): p. 860–866.

149. Lutchen, K.R., et al., Airway constriction pattern is a central component of asthma severity: the role of deep inspirations. *Am J Respir Crit Care Med*, 2001. **164**(2): p. 207–215.

150. Lutchen, K.R., et al., Optimal ventilation waveforms for estimating low-frequency respiratory impedance. *J Appl Physiol*, 1993. **75**(1): p. 478–488.

151. Johnson, A.T., C.S. Lin, and J.N. Hochheimer, Airflow perturbation device for measuring airways resistance of humans and animals. *IEEE Trans Biomed Eng*, 1984. **31**(9): p. 622–626.

152. Bates, J.H., B. Daroczy, and Z. Hantos, A comparison of interrupter and forced oscillation measurements of respiratory resistance in the dog. *J Appl Physiol*, 1992. **72**(1): p. 46–52.

153. Michaelson, E.D., E.D. Grassman, and W.R. Peters, Pulmonary mechanics by spectral analysis of forced random noise. *J Clin Invest*, 1975. **56**(5): p. 1210–1230.

154. Goldman, M.D., et al., Within- and between-day variability of respiratory impedance, using impulse oscillometry in adolescent asthmatics. *Pediatr Pulmonol*, 2002. **34**(4): p. 312–319.

155. Daroczy, B. and Z. Hantos, Generation of optimum pseudorandom signals for respiratory impedance measurements. *Int J Biomed Comput*, 1990. **25**: p. 21–31.

156. Nagels, J., et al., Mechanical properties of lungs and chest wall during spontaneous breathing. *J Appl Physiol*, 1980. **49**(3): p. 408–416.

157. Hantos, Z., et al., Input impedance and peripheral inhomogeneity of dog lungs. *J Appl Physiol*, 1992. **72**(1): p. 168–178.

158. Landser, F.J., J. Clement, and K.P. Van de Woestijne, Normal values of total respiratory resistance and reactance determined by forced oscillations: influence of smoking. *Chest*, 1982. **81**(5): p. 586–591.

159. Peslin, R., C. Duvivier, and C. Gallina, Total respiratory input and transfer impedances in humans. *J Appl Physiol*, 1985. **59**(2): p. 492–501.

160. Bates, J.H., M. Mishima, and Z. Balassy, Measuring the mechanical properties of the lung in vivo with spatial resolution at the acinar level. *Physiol Meas*, 1995. **16**(3): p. 151–159.

161. Mishima, M., Z. Balassy, and J.H. Bates, Assessment of local lung impedance by the alveolar capsule oscillator in dogs: a model analysis. *J Appl Physiol*, 1996. **80**(4): p. 1165–1172.

162. Balassy, Z., M. Mishima, and J.H. Bates, Changes in regional lung impedance after intravenous histamine bolus in dogs: effects of lung volume. *J Appl Physiol*, 1995. **78**(3): p. 875–880.

163. Lutchen, K.R., C.A. Giurdanella, and A.C. Jackson, Inability to separate airway from tissue properties by use of human respiratory input impedance. *J Appl Physiol*, 1990. **68**(6): p. 2403–2412.

164. Finucane, K.E., et al., Resistance of intrathoracic airways of healthy subjects during periodic flow. *J Appl Physiol*, 1975. **38**(3): p. 517–530.

165. Petak, F., et al., Partitioning of pulmonary impedance: modeling vs. alveolar capsule approach. *J Appl Physiol*, 1993. **75**(2): p. 513–521.

166. Hantos, Z., et al., Mechanical impedance of the lung periphery. *J Appl Physiol*, 1997. **83**(5): p. 1595–1601.

167. Bates, J.H. and K.R. Lutchen, The interface between measurement and modeling of peripheral lung mechanics. *Respir Physiol Neurobiol*, 2005. **148**(1–2): p. 153–164.

168. Bates, J.H., et al., Temporal dynamics of acute isovolume bronchoconstriction in the rat. *J Appl Physiol*, 1997. **82**(1): p. 55–62.

169. Lutchen, K.R. and A.C. Jackson, Reliability of parameter estimates from models applied to respiratory impedance data. *J Appl Physiol*, 1987. **62**(2): p. 403–413.

170. Wagers, S., et al., The allergic mouse model of asthma: normal smooth muscle in an abnormal lung? *J Appl Physiol*, 2004. **96**(6): p. 2019–2027.

171. Peslin, R., et al., Frequency response of the chest: modeling and parameter estimation. *J Appl Physiol*, 1975. **39**(4): p. 523–534.

172. Sobh, J.F., et al., Respiratory transfer impedance between 8 and 384 Hz in guinea pigs before and after bronchial challenge. *J Appl Physiol*, 1997. **82**(1): p. 172–181.

173. Aliverti, A., R.L. Dellaca, and A. Pedotti, Transfer impedance of the respiratory system by forced oscillation technique and optoelectronic plethysmography. *Ann Biomed Eng*, 2001. **29**(1): p. 71–82.

174. Dellaca, R.L., et al., Spatial distribution of human respiratory system transfer impedance. *Ann Biomed Eng*, 2003. **31**(2): p. 121–131.

175. Mishima, M., et al., Frequency characteristics of airway and tissue impedances in respiratory diseases. *J Appl Physiol*, 1991. **71**(1): p. 259–270.

176. Tomalak, W., et al., Optimal frequency range to analyze respiratory transfer impedance with six-element model. *J Appl Physiol*, 1993. **75**(6): p. 2656–2664.

177. Lutchen, K.R., Sensitivity analysis of respiratory parameter uncertainties: impact of criterion function form and constraints. *J Appl Physiol*, 1990. **69**(2): p. 766–775.

178. Jackson, A.C., C.A. Giurdanella, and H.L. Dorkin, Density dependence of respiratory system impedances between 5 and 320 Hz in humans. *J Appl Physiol*, 1989. **67**(6): p. 2323–2330.

179. Franken, H., et al., Oscillating flow of a viscous compressible fluid through a rigid tube: a theoretical model. *IEEE Trans Biomed Eng*, 1981. **28**(5): p. 416–420.

180. Benade, A.H., On the propagation of sound waves in a cylindrical conduit. *J Acoust Soc Am*, 1950. **22**: p. 563–564.

181. Bates, J.H., et al., Lung tissue rheology and $1/f$ noise. *Ann Biomed Eng*, 1994. **22**(6): p. 674–681.

182. Hildebrandt, J., Dynamic properties of air-filled excised cat lung determined by liquid plethysmograph. *J Appl Physiol*, 1969. **27**(2): p. 246–250.

183. Fredberg, J.J. and D. Stamenovic, On the imperfect elasticity of lung tissue. *J Appl Physiol*, 1989. **67**(6): p. 2408–2419.

184. Gomes, R.F., et al., Comparative respiratory system mechanics in rodents. *J Appl Physiol*, 2000. **89**(3): p. 908–916.

185. Suki, B., et al., Partitioning of airway and lung tissue properties: comparison of in situ and open-chest conditions. *J Appl Physiol*, 1995. **79**(3): p. 861–869.

186. Fredberg, J.J., et al., Tissue resistance and the contractile state of lung parenchyma. *J Appl Physiol*, 1993. **74**(3): p. 1387–1397.

187. Fust, A., J.H. Bates, and M.S. Ludwig, Mechanical properties of mouse distal lung: in vivo versus in vitro comparison. *Respir Physiol Neurobiol*, 2004. **143**(1): p. 77–86.

188. Sakai, H., et al., Hysteresivity of the lung and tissue strip in the normal rat: effects of heterogeneities. *J Appl Physiol*, 2001. **91**(2): p. 737–747.

189. Hantos, Z., et al., Constant-phase modelling of pulmonary tissue impedance. *Bull Eur Physiopathol Respir*, 1987. **12**: p. 326s.

190. Suki, B., A.L. Barabasi, and K.R. Lutchen, Lung tissue viscoelasticity: a mathematical framework and its molecular basis. *J Appl Physiol*, 1994. **76**(6): p. 2749–2759.

191. Petak, F., et al., Methacholine-induced bronchoconstriction in rats: effects of intravenous vs. aerosol delivery. *J Appl Physiol*, 1997. **82**(5): p. 1479–1487.

192. Bates, J.H., A recruitment model of quasi-linear power-law stress adaptation in lung tissue. *Ann Biomed Eng*, 2007. **35**(7): p. 1165–1174.

193. Bates, J.H., A micromechanical model of lung tissue rheology. *Ann Biomed Eng*, 1998. **26**(4): p. 679–687.

194. Mijailovich, S.M., D. Stamenovic, and J.J. Fredberg, Toward a kinetic theory of connective tissue micromechanics. *J Appl Physiol*, 1993. **74**(2): p. 665–681.

195. Ito, S., et al., Tissue heterogeneity in the mouse lung: effects of elastase treatment. *J Appl Physiol*, 2004. **97**(1): p. 204–212.

196. Suki, B., et al., Partitioning of lung tissue response and inhomogeneous airway constriction at the airway opening. *J Appl Physiol*, 1997. **82**(4): p. 1349–1359.

197. Bellardine Black, C.L., et al., Impact of positive end-expiratory pressure during heterogeneous lung injury: insights from computed tomographic image functional modeling. *Ann Biomed Eng*, 2008. **36**(6): p. 980–991.

198. Kaczka, D.W., et al., Partitioning airway and lung tissue resistances in humans: effects of bronchoconstriction. *J Appl Physiol*, 1997. **82**(5): p. 1531–1541.

199. Hirai, T. and J.H. Bates, Effects of deep inspiration on bronchoconstriction in the rat. *Respir Physiol*, 2001. **127**(2–3): p. 201–215.

200. Bates, J.H. and G.B. Allen, The estimation of lung mechanics parameters in the presence of pathology: a theoretical analysis. *Ann Biomed Eng*, 2006. **34**(3): p. 384–392.

201. Magin, R.L., Fractional calculus in bioengineering, part 3. *Crit Rev Biomed Eng*, 2004. **32**(3–4): p. 195–377.

202. Magin, R.L., Fractional calculus in bioengineering, part 2. *Crit Rev Biomed Eng*, 2004. **32**(2): p. 105–193.

203. Magin, R.L., Fractional calculus in bioengineering. *Crit Rev Biomed Eng*, 2004. **32**(1): p. 1–104.

204. Schetzen, M., *The Volterra and Wiener Theories of Nonlinear Systems*. 1980, New York: Wiley.

205. Suki, B. and J.H. Bates, A nonlinear viscoelastic model of lung tissue mechanics. *J Appl Physiol*, 1991. **71**(3): p. 826–833.

206. Maksym, G.N. and J.H. Bates, Nonparametric block-structured modeling of rat lung mechanics. *Ann Biomed Eng*, 1997. **25**(6): p. 1000–1008.

207. Maksym, G.N., R.E. Kearney, and J.H. Bates, Nonparametric block-structured modeling of lung tissue strip mechanics. *Ann Biomed Eng*, 1998. **26**(2): p. 242–252.

208. Suki, B., Nonlinear phenomena in respiratory mechanical measurements. *J Appl Physiol*, 1993. **74**(5): p. 2574–2584.

209. Suki, B., Q. Zhang, and K.R. Lutchen, Relationship between frequency and amplitude dependence in the lung: a nonlinear block-structured modeling approach. *J Appl Physiol*, 1995. **79**(2): p. 660–671.

210. Zhang, Q., B. Suki, and K.R. Lutchen, Harmonic distortion from nonlinear systems with broadband inputs: applications to lung mechanics. *Ann Biomed Eng*, 1995. **23**(5): p. 672–681.

211. Suki, B., et al., Nonlinearity and harmonic distortion of dog lungs measured by low-frequency forced oscillations. *J Appl Physiol*, 1991. **71**(1): p. 69–75.

212. Suki, B. and K.R. Lutchen, Pseudorandom signals to estimate apparent transfer and coherence functions of nonlinear systems: applications to respiratory mechanics. *IEEE Trans Biomed Eng*, 1992. **39**(11): p. 1142–1151.

213. Hunter, I.W. and M.J. Korenberg, The identification of nonlinear biological systems: Wiener and Hammerstein cascade models. *Biol Cybern*, 1986. **55**(2–3): p. 135–144.

214. Funk, J.R., et al., Linear and quasi-linear viscoelastic characterization of ankle ligaments. *J Biomech Eng*, 2000. **122**(1): p. 15–22.

215. Doehring, T.C., et al., Fractional order viscoelasticity of the aortic valve cusp: an alternative to quasilinear viscoelasticity. *J Biomech Eng*, 2005. **127**(4): p. 700–708.

216. Barabasi, A.L., *Linked. The New Science of Networks*. 2002, Cambridge, MA: Perseus, p. 77.

217. Bak, P., *How Nature Works. The Science of Self-organized Criticality*. 1996, New York, NY: Springer-Verlag, p. 31–37.

218. West, B.J. and M. Schlesinger, On the ubiquity of 1/*f* noise. *Int J Modern Phys*, 1989. **3**: p. 795–819.

219. Bak, P., C. Tang, and K. Wiesenfeld, Self-organized criticality: an explanation of the 1/*f* noise. *Phys Rev Lett*, 1987. **59**(4): p. 381–384.

220. Barabasi, A.L. and R. Albert, Emergence of scaling in random networks. *Science*, 1999. **286**(5439): p. 509–512.

221. Kullmann, L. and J. Kertesz, Preferential growth: exact solution of the time-dependent distributions. *Phys Rev E Stat Nonlin Soft Matter Phys*, 2001. **63**(5 Pt 1): p. 051112.

Index

Printed in the United States
by Baker & Taylor Publisher Services